Estimating Fertilizer Requirements

A Quantitative Approach

Estimating Fertilizer Requirements

A Quantitative Approach

J.D. Colwell
CSIRO Division of Soils
Canberra Laboratories
Acton, Australia

CAB INTERNATIONAL

CAB INTERNATIONAL
Wallingford
Oxon OX10 8DE
UK

Tel: Wallingford (01491) 832111
Telex: 847964 (COMAGG G)
E-mail: cabi@cabi.org
Fax: (01491) 833508

A catalogue record for this book is available from the British Library.

ISBN 0 85198 905 5

Printed and bound in the UK by Biddles Ltd, Guildford.

Contents

Preface

The fertility of soils, as indicated by the yield of the crops they produce and their need for fertilizers, relates to their composition so that it is common for soils to be analysed by "soil testing" laboratories using relatively simple procedures to obtain "soil test values" that can be used to indicate the levels of yields they may produce and their requirements for fertilizers. Way back in the 1950s and 1960s it seemed worthwhile to the author to carry out series of fertilizer experiments to establish relationships between "soil test" analyses and fertilizer requirements for wheat growing regions in southern Australia. These produced the expected relationships but also showed that there was much variation amongst the results of the experiments that was not explained by the tests (Colwell 1963; Colwell and Esdaile 1968). Accordingly a much larger and more ambitious research project was developed, called the National Soil Fertility Project, to establish relationships between soil fertility, as measured with fertilizer experiments, and a wide range of soil and other variables (Hallsworth 1969; Colwell 1977, 1979). This project produced large amounts of data for an extensive region in southern Australia and showed many relationships both with the soil test and other variables. But unfortunately, despite much work, the project also failed to show many of the expected relationships so that for many it ended with disappointment. With the benefit of hindsight and subsequent studies reasons for this failure are now apparent. Thus the effects of many of the factors that affect the results of fertilizer experiments are difficult to demonstrate because they are small and strongly correlated with each other and, most importantly, because at that time appropriate statistical procedures for establishing relationships either had not been developed or were not recognized.

The problems thus demonstrated have stimulated the author over the intervening years to collect and develop the procedures described in this book, that is for research on fertilizer requirements and the relationships between the results of fertilizer experiments and soil and other variables in variable regions. Apart from the economic benefits that such research can produce in regions with severe soil nutrient deficiencies by providing scientific bases for the estimation of economically optimal rates of fertilizer application for crop production, it can also produce environmental benefits in highly developed agricultural regions by providing bases for a conservative use of fertilizers that will both sustain the agriculture and prevent undesirable environmental effects such as the pollution of ground waters by residues of unused fertilizer. The procedures are thus intended to assist research with fertilizers both for regions of low and high soil fertility.

The book is intended for agricultural scientists who are engaged in research on the use of fertilizers with crop production, and for consulting statisticians who may be asked to assist with the interpretation of their data. The scientist often lacks or has forgotten much of the basic mathematics or statistics required to select appropriate procedures and the statistician who, not having had time to properly consider the problems, can easily and unintentionally misguide. There are many traps for the unwary. Accordingly mathematical and statistical procedures are described, endeavouring to indicate their bases and limitations, and how they should be used. Most of the explanations are given in simple terms for the benefit of the ordinary research scientist, assuming only a basic background in mathematics and statistics. The descriptions of some of the mathematical bases for the procedures are however necessarily more technical - these are intended primarily for statisticians and are marked with # so that they may be skipped by the non-specialist reader.

The procedures have been selected and developed on the basis of experience with many soil fertility projects and visits to many research establishments in many countries over many years. They thus represent the result of discussions with many scientists, statisticians and economists as well as of research in various projects. Assistance from all of these sources is gratefully acknowledged. In particular the procedures have grown out of a soil fertility project in Canada with the Ontario Agricultural College (now University of Guelph), in Australia from the aforementioned National Soil Fertility Project, involving CSIRO (Commonwealth Scientific and Industrial Research Organization), State Departments of Agriculture and fertilizer manufacturing companies, and most recently in Brazil from projects with EMBRAPA (Empresa Brasileira de Pesquisa Agropecuária). In particular thanks are due to J. T. Wood of the CSIRO Biometrics Unit, Canberra, who assisted by carefully checking the descriptions of the statistical procedures during their preparation.

Chapter 1

Problems in Estimating Fertilizer Requirements

Usually the simplest and most obvious way to increase crop production in a region with agricultural research is to identify soil nutrient deficiencies and then to determine appropriate applications of fertilizers. The identification of the deficiencies is simple enough, requiring only simple experiments to check for responses in plant growth to trial applications of various nutrients in fertilizers, and the determination of appropriate fertilizer application rates is not much more difficult, at least with simple annual crops such as wheat, requiring only somewhat larger experiments to establish relationships between fertilizer rate and crop yield and then the calculation of the rates that give optimal production. Difficulties soon become apparent however when attempts are made to apply the results of such seemingly straightforward research for the benefit of farmers. Firstly the severity of the deficiencies varies and the fertilizer requirements of crops vary, often markedly, from site to site and from year to year because of variations in the soil, seasonal growing conditions, agricultural practices and so on whereas the results of research are peculiar to the sites and the conditions for the experiments. Thus when a series of fertilizer experiments are carried out in a region, the results will be found to vary with the location of the experiments and, if the experiments are repeated at the same sites, to vary with the year of the experiments. Accordingly if the results of experiments are to be used to guide farmers in the use of fertilizers, they must be adjusted to allow for the effects of variables associated with location and time. Then allowance must be made for the fact that in most situations the cost of applying fertilizers is significant relative to the value of the crop so that economic rates of fertilizer application must be estimated to ensure an optimal economic return from money invested in the use of fertilizers. In other words, fertilizer rates must be estimated that will provide maximum economic returns rather than simply those that, by eliminating nutrient deficiencies, will produce maximum yields. Apart from the economic benefits of such estimates, the consequent conservative use of fertilizers may be expected to prevent undesirable environmental effects of fertilizer use such as the pollution of ground waters by residues of fertilizer not used by the crop. Problems in obtaining estimates can be particularly difficult as when programs of fertilizer application have to be estimated to develop agriculture in regions where there are severe nutrient deficiencies and limited economic resources, when fertilizer rates have to be estimated for the establishment and maintenance of perennial crops such as fruit trees or pastures, or when fertilizer applications affect the value of crops by affecting their quality as well as their yields. Unfortunately, given the many factors that can affect plant growth in a region, often unpredictably as with factors associated with the weather, most estimates of optimal fertilizer rates for the crops of a region based on such research can only be approximate. Nevertheless given data from experiments that represent the variations that occur in a region, best estimates can be obtained on a statistical probability basis which can be expected to be at least better on a long term average basis than estimates obtained on any alternative basis. There are then many worthwhile research challenges for the scientist relating to the use of fertilizers. This book presents mathematical and statistical procedures for such research.

The nature of the problems requiring research relating to the use of fertilizers, and hence the reasons for the development of the procedures to be described, is indicated by way of introduction with the following summaries and questions, grouped under headings corresponding to the chapters that follow. Many of the questions are deceptively simple, simple in that they seem to indicate obvious answers involving the application of standard mathematical and statistical procedures but also deceptive in that they do not indicate the traps for the unwary in the use of these procedures. Accordingly they merit consideration both by scientists and statisticians who are involved in research with fertilizers.

Chapter 2: Exploratory Studies on Fertilizer Requirements

The obvious way to find out whether there is need for research on fertilizer requirements in a region is to carry out some simple experiments that will indicate the relative benefits that can be expected from fertilizer applications for crop production. First questions then concern the choice of designs for simple exploratory experiments on fertilizer requirements and procedures for using the data they may provide to decide whether more comprehensive experiments are warranted. The answers, covered by the material in chapter 2, are mostly fairly obvious but they are important nevertheless because they serve to introduce basic concepts which are used with more difficult subjects in the chapters that follow.

1. What is a suitable design for simple exploratory experiments with fertilizers that will indicate the importance of fertilizer applications for crop production, with about the minimum amount of work consistent with obtaining results that are reasonably reliable and informative?
2. What treatment combinations are needed to show the importance of interaction effects from applications of different nutrients, such that the yield responses to applications of each nutrient are affected by the application of others?
3. What is the purpose of replication, blocks and randomization in experiments?
4. How can the results of fertilizer experiments be represented in the form of regression equations that can be used to calculate estimates of yields for various fertilizer application rates, both for the individual blocks of the experiments and for the experiments considered as a whole?
5. Economically optimal fertilizer rates are defined as those that produce maximum profits. How are regression representations of the results of fertilizer experiments converted to profit functions suitable for the estimation of optimal economic rates, that is by expressing profit from crop production as a function of fertilizer application rate?
6. The calculation of profits from functions that give a mathematical representation of the relationship between fertilizer application rate and crop yield requires allowance for the economic value of the crop and the cost of the fertilizer. In addition, allowance must be made for the delay in time between when an investment is made in the purchase and application of fertilizer and when a return is received after the crop has been grown, harvested and sold. How are profits adjusted by *time discounting* to allow for this time delay, between when money is invested in fertilizer and when an economic return is received?
7. Farmers commonly have many investment opportunities and only limited financial resources. How can provision be made for alternative investment opportunities when calculating the amount of limited financial resources that should be invested in fertilizing,

that is when calculating economically optimal rates of fertilizer application?

8. For cropping systems where several returns are received from a fertilizer application over a period of time, as with successive harvests from perennial crops or the grazing (cropping with animals) of improved pastures, the estimation of optimal rates requires the calculation of the total value of the various returns with adjustments for the times at which they are received. How is the time discounting procedure applied to calculate total economic returns for such systems?

Chapter 3: Optimal Fertilizer Application Rates

Given a profit function $\Pi = f(N)$ expressing profit Π from crop production as a function of fertilizer nutrient application rate N (N does not necessarily indicate nutrient nitrogen), the optimal rate is calculated by solution of the equation $d\Pi/dN = 0$ for the value of N that produces the maximum value for Π, that is for the rate that produces maximum profit. Similarly given a profit function $\Pi = f(N, P, \ldots)$ for several fertilizer nutrients, the simultaneous optimal rates of application of the nutrients N, P, ... is calculated by solution of the simultaneous equations $\partial\Pi/\partial N = \partial\Pi/\partial P = \ldots = 0$ for the rates that produce the maximum.

9. Given yield functions $Y = f(N, P, \ldots)$, such as might be estimated from the data of fertilizer experiments, how are formulas derived for the calculation of optimal nutrient application rates, defined as those rates that produce maximum profit?
10. More specifically, what are formulas for the calculation of optimal rates of application of the nutrients N, P, K given the following fertilizer-yield functions?

$$Y = b_0 + b_1 N^{.5} + b_2 N$$

$$Y = b_0 + b_1 N^{.5} + b_2 P^{.5} + b_3 (NP)^{.5} + b_4 N + b_5 P$$

$$Y = b_0 + b_1 N^{.5} + b_2 P^{.5} + b_3 K^{.5} + b_4 (NP)^{.5} + b_5 (NK)^{.5} + b_6 (PK)^{.5} + b_7 N + b_8 P + b_9 K$$

$$Y = b_0 + b_1 e^{b_2 N}$$

$$Y = b_0 + b_1 \left(N - b_2 - |N - b_2|\right)$$

$$Y = b_0 + b_1 N + b_2 N^2$$

$$\log Y = b_0 + b_1 e^{b_2 N}$$

$$Y = b_0 + b_1 e^{b_2 N} + b_3 N$$

11. A seemingly obvious way to estimate optimal rates is to calculate cost/benefit ratios. Why are these ratios inappropriate for the estimation of optimal fertilizer rates?
12. The simplest calculations of optimal rates are for crops where the only effects of fertilizer application that need to be considered are those that affect a crop yield. However other and more complex effects need to be considered with other types of crop. For example fertilizer can affect the quality and hence the economic value of many crops, as well as their yields, so that allowances must be made for effects on the quality as well as the quantity of crop produce. Also some perennial crops can produce many successive harvests, all of which may be affected by a fertilizer application so that allowance must be made for effects on many yields obtained over a period of perhaps many years. Then for some perennial crops, successive applications of fertilizer are required to establish and maintain the crop so that programs of fertilizer application have to be estimated. How are optimal rates calculated for these situations and types of crop?

Chapter 4: Accuracy of Estimates from Fertilizer-Yield Functions

Yields, profits and optimal fertilizer rates can be calculated with mathematical precision from regression estimates of the functional relationship between fertilizer rate and yield but obviously such values are only estimates, the accuracy of which is limited by the accuracy of the estimate of the relationship from which they are calculated. Various standard statistical procedures are available to indicate the accuracy or reliability of estimates obtained from regressions but these only indicate particular aspects of accuracy and consequently can easily be misused or misinterpreted.

13. What do the following statistical values indicate about the accuracy of estimates obtained from regressions?
    (a) Error variance that is estimated by the residual mean square of regression analyses of variance.
    (b) R^2 or the coefficient of variation.
    (c) R_a^2 or R^2 adjusted for degrees of freedom.
    (d) Tests of significance by analyses of variance.
    (e) Confidence intervals for direct estimates of yield and profit.

(f) Confidence intervals for inverse estimates of rates that produce nominated yields and profits.
(g) Calculated estimates of optimal fertilizer rates.

14. The above measures of accuracy all relate to the error variance that is estimated by analyses of variance of experimental data but there are other major sources of variation that can affect the accuracy of the estimates. What are they?

Chapter 5: Modelling the Fertilizer-Yield Relationship

The use of mathematical and statistical procedures for the interpretation of the results of fertilizer experiments requires the representation of the relationship between fertilizer application rate and crop yield by mathematical functions. Since the form for these functions is not defined by theory, an empirical model form must be chosen, as for example from the commonly used functions

$$Y = b_0 + b_1 N^{.5} + b_2 N$$

$$Y = b_0 + b_1 \, e^{b_2 N}$$

$$Y = b_0 + b_1 \left(N - b_2 - \left| N - b_2 \right| \right)$$

$$Y = b_0 + b_1 N + b_2 N^2$$

for a single nutrient relationship, $Y = f(N)$. The selection is important because alternatives can give very similar R^2 values and tests of significance, but nevertheless give very different representations of the relationship, and very different estimations of optimal fertilizer rates.

15. What bases should be used to select models for regressions to estimate the fertilizer-yield relationship from the data provided by fertilizer experiments?
16. What are the good and bad features of the above alternative models?
17. How are models for multiple nutrient functions derived from a single nutrient model?

Chapter 6: Analysis of Polynomial Functions

A major reason for using polynomial models on the square root scale to represent the fertilizer-yield relationship is that polynomials provide convenient analyses of variance of experimental data into independent or orthogonal components. The separation of treatment effects into orthogonal components is important because it provides a means for estimating regression relationships between the results of experiments and experiment site variables that can then be used collectively to predict corresponding experimental results and hence to estimate fertilizer requirements for sites in a variable region. To appreciate how this is possible with an appropriate choice of polynomial regressions for the fertilizer-yield relationship requires an understanding of the nature of orthogonal analyses of variance, orthogonal polynomials and, for the statistician, an understanding of the mathematics on which they are based.

18. What are orthogonal polynomials and how are they calculated?
19. What is the relationship between the coefficients of orthogonal trends in orthogonal polynomial regressions such as $Y = p_0\xi_0 + p_1\xi_1 + p_2\xi_2 + p_3\xi_3$ and the components of a detailed analysis of variance for corresponding non-orthogonal polynomial regressions such as $Y = b_0 + b_1X + b_2X^2 + b_3X^3$ or, on the square root scale, $Y = b_0 + b_1N^{.5} + b_2N + b_3N^{1.5}$?
20. What are the mathematical relationships between the coefficients b_0, b_1, ... of non-orthogonal polynomial regressions and the coefficients p_0, p_1, ... of the same regressions in the form of orthogonal polynomial regressions?
21. Which coefficients of non-orthogonal polynomial regressions are directly proportional to coefficients of orthogonal polynomial regressions and consequently suitable for establishing predictive relationships between the results of fertilizer experiments and site variables?

Chapter 7: Designs for Fertilizer Experiments

The estimation of optimal fertilizer application rates from the data provided by fertilizer experiments requires the estimation of the relationship between fertilizer application rate and crop yield. Such relationships can be estimated efficiently from the data provided by suitably designed fertilizer experiments and orthogonal polynomial regressions for the relationship can be used to indicate appropriate designs.

22. What are the components of polynomial yield functions that have to be estimated from the data of fertilizer experiments and how are treatments chosen to best estimate these components?
23. What are suitable designs and treatments for estimating the following polynomial yield functions?

$$Y = b_0 + b_1N^{.5} + b_2N$$

$$Y = b_0 + b_1N^{.5} + b_2P^{.5} + b_3(NP)^{.5} + b_4N + b_5P$$

$$Y = b_0 + b_1 N^{.5} + b_2 P^{.5} + b_3 K^{.5} + b_4 (NP)^{.5} + b_5 (NK)^{.5} + b_6 (PK)^{.5}$$
$$+ b_7 N + b_8 P + b_9 K$$

$$Y = b_0 + b_1 N^{.5} + b_2 P^{.5} + b_3 K^{.5} + b_4 S^{.5} + b_5 (NP)^{.5} + b_6 (NK)^{.5} + b_7 (NS)^{.5}$$
$$+ b_8 (PK)^{.5} + b_9 (PS)^{.5} + b_{10} (KS)^{.5} + b_{11} N + b_{12} P + b_{13} K + b_{14} S$$

24. Central composite type designs are recommended for the estimation of response surfaces in standard books on experimental design but this type of design can prove unsatisfactory for the estimation of fertilizer-yield functions. What are the reasons for not using central composite "response surface" designs for the estimation of the fertilizer-yield relationship and hence of optimal fertilizer rates?
25. In regions with severe soil nutrient deficiencies estimates of optimal programs of fertilizer application are required that will raise soil fertility to optimal levels over a period of time with the accumulation of residues from fertilizer applications. What designs can be used to provide data suitable for estimating such programs?

Chapter 8: General Soil Fertility Models

When a series of fertilizer experiments is carried out in a variable region, results vary for the different experiments due to variations in variables associated with the experiments, such as soil nutrient levels, kind of soil at the experiment site, the weather that prevailed for the experiments and the agricultural practices used in the establishment and management of the experiments. If relationships can be established between the results of such experiments and these variables, called *site variables*, they can be used both to explain the variations amongst the results of the experiments and to provide a basis for estimating corresponding results for other sites and other years in the region. A set of equations that can be used to obtain such estimates is called a *general soil fertility model*. The statistical estimation of such models requires the representation of the results of experiments by variables corresponding to orthogonal components of a detailed analysis of variance for each experiment called *yield variables*.

26. What is meant by orthogonal components of the yield variance and why must these components be used to define yield variables for studies on the relationships between the results of fertilizer experiments and site variables?
27. How are suitable variable data for such studies obtained from experimental data?
28. How are the relationships used to predict yield functions?

29. What are variable parameter functions and how can such functions be derived from general soil fertility models?
30. If data from fertilizer experiments are used to estimate Mitscherlich functions in the form $Y = a[1 - e^{c(N+b)}]$ where N is nutrient application rate, biological meanings can be associated with the model parameters, a = maximum attainable yield with fertilizer, b = soil nutrient content and c = an efficiency or scaling factor. Furthermore regression relationships can be established between these parameter values and site variables as suggested by these biological meanings and the regressions then used to estimate values for the model parameters $\hat{a}$, $\hat{b}$ and $\hat{c}$. Given a set of regression estimates $\hat{a}$, $\hat{b}$ and $\hat{c}$ of the parameters for a site with a particular set of site variable values, why should they *not* be used to estimate a function for that site, as in $Y = \hat{a}[1 - e^{\hat{c}(N+\hat{b})}]$? Similarly, why are otherwise very satisfactory models for the fertilizer-yield relationship unsuitable for the development of general models?

Chapter 9: Development and Use of General Soil Fertility Models

The procedure for developing general soil fertility models is essentially that for the development of regressions, the distinctive feature being that the dependent variables, called yield variables, are especially chosen to correspond to orthogonal components of regressions for fertilizer-yield functions. Yield variable values for such regressions are obtained by fertilizer experiments and corresponding regressor variables, called site variables, by measurements and observations at the sites of the experiments. Data values for some site variables such as chemical analyses of the surface soil horizon can be obtained by direct measurement but others require computation from several site measurements. Thus when the distribution of chemical analyses down soil profiles affects experimental results, profiles must be sampled and the data used to compute suitable site variables. Similar computations are required to obtain site variables for weather effects. Other site variables such as kind of soil or location in a climatic zone cannot be quantified and must be represented by dummy variables.

31. How can soil profile features be represented in a suitable form for use as site variables in regressions?
32. How are dummy variables defined for non-quantifiable site variables such as the kind of soil at experiment sites?
33. How are dummy variables used in regressions for general soil fertility models?
34. What are suitable site variables for weather effects?

35. When general models have been established relating the results of experiments to site variables there are questions concerning the accuracy and usefulness of the estimates that they can provide. How can the accuracy of estimates obtained for general models be indicated?
36. What are the major sources of error that affect the accuracy of estimates from general models?
37. Finally, given all of the difficulties and problems raised by these questions, a last question might be: What is wrong with the relatively simple procedures that have been used in the past to interpret the results of fertilizer experiments and to estimate fertilizer requirements?

Chapter 2

Exploratory Studies on Fertilizer Requirements

The obvious way to obtain information about the importance of fertilizers for crop production and the benefits that may be expected from research on their use is to carry out simple experiments with the aim of showing the effects that fertilizer applications will have on crop yields and, most importantly, the economic benefits that farmers are likely to gain from the use of fertilizers. The experiments can be quite small and simple, consisting for example of a few test applications of fertilizer in strips across a field that can be used to obtain direct comparisons of crop growth with and without fertilizer. However, unduly simple experiments can produce misleading impressions about fertilizer requirements. For example the application of nitrogen fertilizer in a strip across a field will usually produce a greener and more vigorous crop giving the impression that the fertilizer has had a beneficial effect whereas the actual effect on the marketable product may be negligible or even negative, as with sugarcane when the effect of nitrogen fertilizer may be to increase the production of cane but to decrease its sugar content or with wheat grown under semiarid conditions when the effect may be to increase the production of straw but to decrease the amount of harvestable grain. Moreover apparent treatment effects in a simple experiment may be confusing or misleading due to the effects of other irrelevant or unrecognized sources of variation that are likely to affect the results of any experiment, called *experimental error*. Consequently even the simplest of what may be called an exploratory, demonstration or diagnostic experiment should be carefully designed so that (a) it will produce the required information and (b) treatment effects can be distinguished from experimental error effects. Figure 2.1 illustrates such a simple

	Treatment No.	Code	Plot yield
Block 1	2	01	2917
	3	10	1450
	1	00	1375
	4	11	3283
Block 2	4	11	3660
	2	01	3184
	3	10	1253
	1	00	1125
Block 3	3	10	1305
	2	01	2830
	4	11	3317
	1	00	1461

Fig. 2.1 Layout for a simple experiment designed to investigate the importance of N and P fertilizers for crop production. The treatments that produce the plot yields are identified by the treatment codes defined in Table 2.1.

experiment with a design that has been chosen (a) to produce information on the effects of applications of nitrogen and phosphorus fertilizer on the yield of a crop, each in the presence and absence of the other, and (b) to show error effects as well as treatment effects, with about a minimal amount of work combined with careful management. The features of the design and the interpretation of the data (plot yield) that it has produced with this example experiment are described in some detail in this chapter in order to introduce terms, notation and basic concepts which are used in the following chapters.

2.1 Features of designs for fertilizer experiments

Ideally experiments are designed so that their treatments will produce pertinent information with an acceptable level of accuracy and so that they are efficient in the sense that effort is not wasted obtaining unwanted or meaningless information, or results with an unnecessarily high level of precision. Suitable designs are thus chosen by considerations of:

1. treatments that will produce data for specified types of information and
2. the likely levels and nature of error variations that may affect the data and hence the accuracy of the information they provide.

Such considerations led to the choice of the design for the example experiment. Firstly treatments were chosen that consist of *factorial* combinations of rates of application of the fertilizer nutrients N and P, that is all combinations of the application rates 0 and 100 kg N/ha with 0 and 50 kg P/ha. Secondly all of these treatments were replicated in each of three blocks to provide for possible error effects. The same considerations lead to similar types of design for larger and more comprehensive experiments, as described in chapter 7.

Treatment and yield codes

The treatments for the example experiment are given in Table 2.1 together with three alternative treatment coding systems that are commonly used with factorial designs for experiments. Coding system 1, with the codes 0 and 1 corresponding to the two nutrient rates for each nutrient, is generally preferred because of its simplicity but it does require knowledge of the treatment factors (application rates of N and P) that correspond to each integer. The coding system 2 is more explicit in this respect but is unnecessarily cumbersome since only the nutrient subscripts vary as indicated more simply with the system 1. The coding system 3 is given because it has been used in many standard reference

Table 2.1 Treatments and treatment codes for the example experiment in Fig. 2.1.

Treatment	Nutrient rates		Treatment codes			Data
number	kg N/ha	kg P/ha	1	2	3	code
1	0	0	00	N_0P_0	(1)	Y_{00}
2	0	50	01	N_0P_1	p	Y_{01}
3	100	0	10	N_1P_0	n	Y_{10}
4	100	50	11	N_1P_1	np	Y_{11}

books on experimental design for factorial designs. This system is only suitable however for designs where treatments can be represented bythe presence or absence of factors, such as of 100 kg N/ha and 50 kg P/ha for this example. It is consequently not suitable for designs for fertilizer experiments which require three or more nutrient treatment rates in order to estimate the typically curved relationships between yield and application rates.

The yield data produced by the treatments is also shown in code form, Y_{00}, Y_{01}, Y_{10} and Y_{11}, to indicate in a general form data produced by treatments corresponding with the subscripts. This type of coding is convenient for indicating computational procedures for data interpretation and analysis as illustrated below.

Factorial combination of nutrient treatment rates

The basic objective with the example experiment is to obtain information about crop yield responses to applications of N and P fertilizers, each in the presence and absence of the other, and then to use this information to indicate the economic importance of fertilizers as a guide for future studies. On the basis of these objectives and of knowledge about effects of applications of N and P in fertilizers in other regions, the application rates 100 kg N/ha and 50 kg P/ha were chosen as treatments that could be expected to correct any likely N or P soil nutrient deficiencies and the zero application rates 0 kg N/ha and 0 kg P/ha were chosen to give bases for comparison or contrast that show the effects of the correction of any nutrient deficiencies. The treatment combinations were chosen similarly to give contrasts between the yield response to each nutrient in the presence and absence of the other, and hence to show the N × P *interaction* effect. Thus denoting the yield data produced by the treatments as Y_{00}, Y_{01}, Y_{10} and Y_{11}, as in Table 2.1, the following differences or contrasts show the indicated effects:

$Y_{01} - Y_{00}$: response to 50 kg P/ha in the presence of 0 kg N/ha,

$Y_{11} - Y_{10}$: response to 50 kg P/ha in the presence of 100 kg N/ha,

$Y_{10} - Y_{00}$: response to 100 kg N/ha in the presence of 0 kg P/ha and

$Y_{11} - Y_{01}$: response to 100 kg N/ha in the presence of 50 kg P/ha.

Given these responses, the N × P interaction effect is shown by the differences between the responses to each nutrient in the presence and absence of the other:

$$(Y_{11} - Y_{10}) - (Y_{01} - Y_{00})$$

for the response to P in the presence and absence of N, and identically,

$$(Y_{11} - Y_{01}) - (Y_{10} - Y_{00})$$

for the response to N in the presence and absence of P.

The combination of treatment rates chosen to give these data contrasts is termed a *factorial combination* and the design a *factorial design*. For this very simple example design there are $2 \times 2 = 2^2$ factorial treatment combinations so that the design may be described as being a 2×2 or a 2^2 factorial. The same considerations in the choice of treatments to produce contrasts for the determination of specific effects and their interaction effects may be applied for more complex quadratic, cubic, etc. effects, as required to represent the effects produced with fertilizer applications at more than two levels. Thus more generally, if a design consists of factorial combinations of n factors with the respective number of levels $t_1, t_2, \ldots, t_n$ for each factor, then it is described as being a $t_1 \times t_2 \times \ldots \times t_n$ factorial design. If the number of treatment levels for the factors are the same this may be condensed to a t^n type description as for example with $3 \times 3 \times 3 = 3^3$, $4 \times 4 = 4^2$ or $2 \times 2 \times 2 \times 2 = 2^4$.

Replication for accuracy

The data produced by experiments are affected, seemingly inevitably, by unknown or uncontrolled factors that produce so-called error effects and consequently designs must be developed to allow the estimation of treatment effects from the data with an acceptable level of accuracy, despite these error effects. If the error effects can be assumed to occur randomly amongst the data and to be positive and negative with a mean value of zero, the obvious way to improve the accuracy of estimates from experiments is to replicate the treatments. Error effects will then tend to cancel out when data are averaged to estimate the treatment effects. Thus for the example experiment, the four treatments in Table 2.1 have been replicated in three blocks (Fig. 2.1) with the expectation that the accuracy of the estimates of treatment effects from the combined data will thereby be improved. The replication also provides a means for estimating the magnitude of the error effects in the form of an error variance that can be used to judge the accuracy of the estimated treatment effects, as with statistical tests of significance and confidence levels.

The simple replication of treatments is however only one way, and not necessarily the best way, to increase the accuracy of experimental results. Another, often more efficient, way is to use the factorial combinations of the treatment rates in factorial designs to provide a form of replication. Thus for the example design (Table 2.1) if there is no N × P interaction effect such that applications of N or P have independent effects on crop yields, the pairs of treatments for each nutrient are in effect replicates and average or mean treatment effects are estimated from the combined data by

$$\text{N effect} = \left(Y_{10} - Y_{00} + Y_{11} - Y_{01}\right) \div 2$$

$$\text{P effect} = \left(Y_{01} - Y_{00} + Y_{11} - Y_{10}\right) \div 2$$

If there is an interaction effect, this effect can be separated mathematically from the average effects of the individual nutrients so that the accuracy of the estimates is still improved by the factorial combination of the treatment rates. The factorial combination of treatments thus provides a type of replication often described as the *inbuilt replication* of factorial designs. This type of replication is particularly useful for larger experiments because with these, it can completely replace the need for replication by repetition and thus reduce very substantially the size of experiments. Decreasing the size of experiments in this way, as well as reducing work requirements, can increase the accuracy of experiment results by allowing the use of smaller experiment sites which, because they are smaller, are likely to be less variable.

Blocks as dummy treatments

Ideally field experiments are located on uniform sites to avoid error effects caused by variations of factors such as the soil, previous cultivation history or drainage within the experiment sites. In some situations however it is difficult or impossible to choose sites that will not include obvious potential sources of within site variation, as when experiments must extend over several small fields in regions of intensive agriculture or over adjacent terraces in regions with a mountainous terrain. Blocking in experimental designs provides a means to allow for the effects of such within site variations and hence to improve the accuracy of estimates of treatment effects by reducing the unexplained or residual "error" variance in analyses of variance. This is done by allocating treatments to blocks in such a way that the variance due to block differences can be separated in analyses of variance. Thus if the example experiment had to be located in a mountainous region where farming was on small terraces, the three blocks of four plots could have been located in three adjacent terraces to allow for the within site variation due to differences between the terraces. Blocking may also increase the accuracy of experiments even when experiment sites appear to be uniform since the blocks may perchance be located to coincide with some unperceived within site variations. Standard experimental designs consequently have treatments grouped into blocks that are as small as possible consistent with the mathematical requirements for the separation of treatment and block effects. For small experiments blocking is accomplished simply by repeating the treatments in each block as in Fig. 2.1, giving replication in blocks, but for larger designs more complex mathematical procedures are used to select groups of treatments for the blocks in such a way that the blocks are smaller than would be required by simple replication in blocks.

The use of blocks to allow for variations within experiment sites may be regarded as equivalent to the use of blocks as treatments, the "block treatments" being applied by the allocation of plots to the blocks and the "treatment" effects being produced by the variations in soil, etc. amongst the blocks. The actual factors that produce the "block treatment" effects need not be known for the representation of their effects in regressions since *dummy variables* can be used for the "block treatments". Orthogonal polynomials as given in standard statistical tables (for example Fisher and Yates 1963) provide convenient values for such variables. Thus for the three blocks of the example, "block treatments" are represented by the dummy variables L_b and Q_b in Table 2.2 with values −1, 0, +1 and +1, −2, +1 according to the location of the plots in blocks 1, 2 or 3. The variable L_b for linear orthogonal trend serves to represent any uniform increase or decrease in yield levels associated with location from block 1 to 3 across the experiment site and similarly the variable Q_b for quadratic orthogonal trend serves to represent any block differences such that the yield level in block 2 is higher or lower than the level in the adjoining blocks 1 and 3, after allowing for the linear trend.

Randomization

The nature of experimental designs and the effects of treatments are easier to comprehend if the treatments and the data they produce are tabulated in an orderly fashion as in Table 2.1 and below in Table 2.2. Unfortunately such orderly arrangements can mislead experimenters into using correspondingly orderly arrangements when establishing experiments rather than a random and, in this sense, disorderly arrangement as in Fig. 2.1. Any such orderly arrangement of treatments for the layout of experiments must be avoided because the order may produce bias favouring the estimation of particular treatment effects. For example if an experimental layout was chosen to correspond to the orderly arrangement of treatments in Table 2.2, it could produce a bias

Table 2.2 An orderly arrangement of treatments and yield data for the example experiment in Fig. 2.1. Plot numbers correspond to those used for the layout in the experiment site. Treatments include dummy variables L_b and Q_b for block location of the plots.

Plot	Block	Code	Treatments				Yield
			kg N/ha	kg P/ha	L_b	Q_b	kg/ha
3	1	000	0	0	-1	+1	1375
1	1	010	0	50	-1	+1	2917
2	1	100	100	0	-1	+1	1450
4	1	110	100	50	-1	+1	3283
8	2	001	0	0	0	-2	1125
6	2	011	0	50	0	-2	3184
7	2	101	100	0	0	-2	1253
5	2	111	100	50	0	-2	3660
12	3	002	0	0	+1	+1	1461
10	3	012	0	50	+1	+1	2830
9	3	102	100	0	+1	+1	1305
11	3	112	100	50	+1	+1	3317

favouring the measurement of yield responses to the P treatments more than of corresponding responses to the N treatments because it gives more side by side comparisons for the effects of the 0 and 50 kg P/ha treatments than of the 0 and 100 kg N/ha treatments. For this reason, treatments should be allocated randomly to plots using random numbers as given in standard statistical tables or as generated by a suitable computer algorithm. Similarly, since any particular randomization may favour a particular treatment effect because of a chance arrangement of treatments, different randomizations should be used for each replicate in blocks and for each experiment when there is a series of experiments. For the example the three randomizations 2, 3, 1, 4; 4, 2, 3, 1 and 3, 2, 4, 1 of the treatment numbers 1, 2, 3, 4 of Table 2.1 were used to obtain the layout of treatments shown in Fig. 2.1.

2.2 Interpretation of experimental data

Data inspection

The first thing that an experimenter should do at the conclusion of an experiment is to *inspect the data* to obtain a first impression of the results of the experiment. Although this injunction may seem ridiculously obvious it is nevertheless necessary because of the ever present temptation to simply feed data into a computer and then to accept what comes back out in a neatly printed and authoritative form, as though the computer is some sort of intelligent and infallible robot. Although the computer processing of data is immensely valuable in that it can provide a means for practically all of the interpretations of the data obtained from experiments, it is important to understand the nature and limitations of the processing in order to ensure that appropriate procedures are used and appropriate conclusions drawn from the results. A first step then, at the conclusion of an experiment, is to carefully inspect the data for treatment effects to provide a basis for checking and interpreting the results produced by the data processing. Data inspection is also needed to detect erroneous data, produced for example by an error when recording or copying. Such errors may greatly affect the results of any processing.

Data inspections for treatment effects require orderly tabulations, as in Table 2.3 for the example experiment. Appropriate tabulations such as this are often indicated for standard designs by the tabulations that used to be used for statistical analyses of variance with desk calculators, as described in textbooks of the precomputer era. For the example, comparisons of rows and columns in Table 2.3, and of their totals and means, show the effects of the treatments, the variations between the blocks, and the error aberrations amongst the data. Thus an inspection of the data shows a large yield response to the P treatment, a small response to the N treatment and a small positive N × P interaction effect such that responses to N or P are greater, each in the presence of the other. Specifically:

1. the means for the P treatments show an increase from 1328 to 3199 due to the application of P,
2. the means for the N treatments show an increase from 2149 to 2378 due the application of N,
3. an interaction effect is shown by comparing means for the yield response to P for the alternative N treatment rates, comparing $2977 - 1320 = 1657$ when $N = 0$ with $3420 - 1336 = 2084$ when $N = 100$, and the effect is described as a *positive interaction* because the response is greater when the N rate is greater,

Table 2.3 Tabulations of data for the experiment in Fig. 2.1 to show treatment and error effects.

Treatments		Blocks			Total	Mean
kg N/ha	kg P/ha	1	2	3		
0	0	1375	1125	1461	3961	1320
0	50	2917	3184	2830	8931	2977
100	0	1450	1253	1305	4008	1336
100	50	3283	3660	3317	10260	3420
	Total	9025	9222	8913	27160	
	Mean	2256	2306	2228		2263

kg N/ha	kg P/ha		Mean
	0	50	
0	1320	2977	2149
100	1336	3420	2378
Mean	1328	3199	2236

4. alternatively the positive interaction effect is shown by comparing the response to N with the alternative P treatment rates, $1336 - 1320 = 16$ with $3420 - 2977 = 443$,
5. the effect of location in blocks is shown by the block means 2256, 2306 and 2228.

Inspection may also indicate error aberrations such that data for particular treatment effects vary amongst the replicates. Error effects for statistical analyses of variance are supposed to vary randomly, producing both positive and negative effects on the data and having a mean effect of zero. As such error effects for statistical purposes should not represent the effects of some identifiable error or mistake that has occurred during the experiment and which can be either corrected or avoided by rejecting the affected data. If for example a plot was damaged during the course of an experiment, the effect would not be random and the consequent error should either be corrected, as by estimating a yield value from the undamaged portion of the plot, or excluded by simply rejecting that plot datum. In such cases missing data procedures are used for analyses of variance and estimations of treatment effects. For this example it is to be assumed that experience with similar experiments indicates that the error effects are normal for this type of experiment and that there is no identifiable cause of error that would justify any correction or rejection of data.

Analysis of variance

Analyses of variance of experimental data, as the name implies, separate or analyse data variations as represented by their variance, into components, the analysis procedures being chosen so that the components correspond with possible treatment, block and error effects and thus provide bases for interpreting the results of experiments. More specifically the variation of a set of data is measured by the sum of squares of deviations of the data about their mean, called the total sum of squares (TSS), computed by

$$\mathrm{TSS} = \sum_{i=1}^{n}\left(Y_i - \overline{Y}\right)^2 = \sum_{i=1}^{n} Y_i^2 - n\overline{Y}^2 \qquad (2.1)$$

where the Y_i are n yield data and $\overline{Y}$ is the data mean. Then the analysis of variance is obtained by separating this total sum of squares into components. The data variance is calculated by dividing the total sum of squares by its degrees of freedom, Variance = $\mathrm{TSS}/(n-1)$, and the components of the TSS separated by the analysis are also divided by their respective degrees of freedom to obtain *mean squares* for direct comparisons and tests of significance.

The standard analysis of variance for the example data is illustrated in Table 2.4 together with the calculation procedure used with desk calculators in the precomputer era. Firstly the total sum of squares, TSS = 11,025,995, is computed by (2.1) and then component sums of squares corresponding to the effects noted above for the data inspection are computed as detailed in the lower part of the table. The numerical values for the sums of squares are large and cumbersome so that for convenience of comparison they are expressed as percentages of the total in column 4 of the analysis table. Mean squares are obtained by dividing the sums of squares by their respective degrees of freedom and F ratios for tests of significance are obtained by dividing these mean squares by the residual mean square. Traditionally levels of significance indicated by the magnitudes of these F ratios and their respective degrees of freedom are given the star ratings, * for $p<0.05$, ** for $p<0.01$ and *** for $p<0.001$ and a value with a probability $p>0.05$ is described as being not significant. For the example, F values with 1 and 6 degrees of freedom for these significance ratings are respectively 5.99, 13.74 and 35.51 so that only the P effect with F=280.39 gets a star rating. The other F ratios have values less than 5.99 so that they are described as being not significant in the sense that there is a probability greater than 1 in 20 (5%) that the effect could be simply due to variations amongst the data produced by chance error

Table 2.4 Analysis of variance for the example experiment. Data values for the calculations are taken from Table 2.3.

Source	Degrees of freedom	Sum of squares	% Sum of squares	Mean square	F ratio
Blocks	2	12236	0.11	6118	0.16
N	1	157781	1.43	157781	4.22
P	1	10494440	95.17	10494440	280.39***
N × P	1	136960	1.24	136960	3.66
Residue	6	224577	1.04	37430	
Total	11	11025995	100.00		

Sum	Calculation
Correction	$C = n\overline{Y}^2 = \dfrac{(1375 + 1125 + \cdots + 3317)^2}{12} = 61472133$
Block	$\dfrac{9025^2 + 9222^2 + 8913^2}{4} - C = 12236$
N	$\dfrac{12892^2 + 14268^2}{6} - C = 157781$
P	$\dfrac{7969^2 + 19191^2}{6} - C = 1049440$
N × P	$\dfrac{3961^2 + 8931^2 + 10260^2}{3} - \text{N effect} - \text{P effect} - \text{C} = 136961$
Total	$1375^2 + 1125^2 + \cdots + 3317^2 - C = 11025995$

effects. It is important to understand in this regard that the "not significant" rating does not imply that effects are nil or even that they are probably nil. In particular the F= 4.22 value for the N effect is not far below the critical value 5.99 for a * rating so that it is worth noting, for comparison with the results of other similar experiments or as a pointer to something that may be worth checking with further research.

Regression estimates of treatment effects

Analyses of variance do not estimate treatment effects, nor even whether they are positive or negative, but merely indicate their relative magnitudes and statistical significance. Rather the nature of the effects is seen from inspections of the data or more precisely by estimating the magnitude of the effects by a statistical estimation procedure, as by comparison of the means of replicate data for the various treatments. For fertilizer experiments where the treatments are numerical values for quantitative variables, that is for nutrient application rates, statistical estimates can be calculated from regression equations that have been estimated from all of the data for each experiment. Regression equations obtained in this way are useful because:

1. They can provide single algebraic representations of all of the results of each experiment which can be used to calculate specific treatment effects on both crop yield and economic return for particular economic situations.
2. They can be used to calculate corresponding values for effects of fertilizer application rates other than those actually used as treatments and hence to estimate optimal rates.
3. They can be used to calculate *yield variables* corresponding to the components of variance separated in analyses of variance. These special variables, which are described in following chapters, are useful for studies on relationships between the results of experiments and site variables in variable regions.

The example experiment is used to introduce these uses for regressions that have been estimated from the experimental data, although with such a simple experiment, studies would usually be restricted to an inspection of the data to supplement the analysis of variance. Since there are only two treatment rates for each of the fertilizer nutrients, estimates of yields and profits for rates other than the treatment rates involve interpolation between yield and nutrient application rate without information concerning the form of curvature for the relationship. Such interpolated estimates may be very misleading.

Regression models

Many alternative regression forms or models can be estimated from experimental data and hopefully the model chosen will give a good representation of the form of the biological causal relationship that has produced the treatment effects. For the present example the causal relationship is the effects on crop yield of N and P nutrient application rates and location in the blocks of the experiment site. If a model is required to only represent the effects of the two treatment rates for each nutrient and of location in the three blocks, mathematically simple models such as

$$Y = b_0 + b_1N + b_2P + b_3NP + b_4L_b + b_5Q_b \quad (2.2)$$

and

$$Y = a_0 + a_1N + a_2P + a_3L_b + a_4Q_b \quad (2.3)$$

can be used, where N and P are application rates for the nutrients, and L_b and Q_b are the dummy variables for block effects, all as listed under treatments in Table 2.2. Regression estimates of these models obtained from the example data are:

$$Y = 1320.3 + 0.1567N + 33.13P + 0.0855NP - 14.0L_b - 21.0Q_b \quad (2.4)$$

and

$$Y = 1214 + 2.293N + 37.41P - 14.0L_b - 21.0Q_b \quad (2.5)$$

The first of these regressions gives the better representation of the experimental results because it provides for a possible N × P nutrient interaction effect with the variable term NP. The second is also important however, because it can be used in conjunction with the first to calculate the sum of squares attributable to the interaction effect for the analysis of variance in Table 2.4 and hence to calculate the statistical significance of the interaction effect. Thus there is a mathematical and direct relationship between the coefficients 2.293 and 37.41 for the variables N and P in (2.5) and the respective sums of squares for N and P effects and between the coefficient 0.0855 for the NP variable in (2.4) and the sum of squares for the N × P interaction as will be described in chapter 6. Regression coefficients that have such direct relationships with components of an analysis of variance are important because they can be used as data values for variables, to be called *yield variables*, to establish relationships between the results of series of experiments in a variable region and site variables, as will be described in chapters 8 and 9 for the development and use of *general soil fertility models*.

The magnitudes of the block effects on yield can be calculated by substituting the values for these dummy variables, listed in Table 2.2, in the equations. The coefficient values –14.0 and –21.0 for the block variables L_b and Q_b also relate directly to components of the analysis of variance, corresponding to the linear and quadratic trend effects in Table 2.4. Ordinarily such estimates for individual blocks will be of little interest however, the important estimates being those for the experiment considered as a whole. Since the mean values for the L_b and

Q_b are zero, equations representing treatment effects for the whole experiment site are obtained by substituting $L_b = Q_b = 0$ or equivalently, by simply omitting the block terms from the regressions, to obtain

$$Y = 1320.3 + 0.1567N + 33.13P + 0.0855NP \qquad (2.6)$$

from the regression (2.4).

Alternative regression models

Corresponding representations of treatment effects can be obtained using other regression models. In particular the models

$$Y = b_0 + b_1 N^{.5} + b_2 P^{.5} + b_3 (NP)^{.5} + b_4 L_b + b_5 Q_b \qquad (2.7)$$

and

$$Y = a_0 + a_1 N^{.5} + a_2 P^{.5} + a_3 L_b + a_4 Q_b \qquad (2.8)$$

might be used because of an expected curved relationship between yield and nutrient application rate, giving the alternative regression representations of the experimental results

$$Y = 1320.3 + 1.567N^{.5} + 234.3P^{.5} + 6.046(NP)^{.5} - 14.0L_b - 21.0Q_b \qquad (2.9)$$

and

$$Y = 1214 + 22.93N^{.5} + 264.5P^{.5} - 14.0L_b - 21.0Q_b \qquad (2.10)$$

These regressions also relate directly to the analysis of variance in Table 2.4 and give identical estimates of treatment effects to the regressions (2.4) and (2.5). However, the alternative regressions give different interpolated estimates of yield for nutrient application rates other than the treatment rates of 0 and 100 kg N/ha and 0 and 50 kg P/ha and in the absence of data to indicate the form of curvature for the relationship, it is simply a matter of opinion which estimates are the best. When experiments have only two treatment rate levels, they cannot provide any information about the form of the relationship for other application rates so that any estimate involving either interpolation or extrapolation must be suspect - this of course is the reason for the use of more than two treatment levels for other than exploratory experiments. The models (2.2) and (2.3) are preferred for experiments with two treatment levels only because they are mathematically simpler and not because interpolated or extrapolated estimates from regressions for these models are necessarily likely to be better than those obtained with alternatives such as (2.7) and (2.8).

2.3 The practical significance of experimental results

Practical questions

A hazard with scientific research is that the scientist, in the process of applying sophisticated procedures for the interpretation of experimental data and producing esoteric publications, is liable to forget to answer in direct and simple terms the questions which led to the research in the first place. Such questions can be expected to arise again in the practical application of the results and if answers are not volunteered, then it is likely that there will be no practical application or even an incorrect application. A safeguard in this respect would be for the scientist to have to explain the results and their practical implications directly to the people who are expected to benefit from the research. A more workable procedure is for the scientist to formulate practical questions and answers relating to the application of the results, both to assist in their reporting for the benefit of users and to indicate directions for further research. Practical questions relating to research with fertilizers are concerned with improvements in agricultural production by the use of fertilizers and, in particular, with economic benefits. Thus for the example experiment the following questions might be considered by the scientist:

1. *Are the yield responses to the fertilizer treatments large enough to be of practical importance and hence to warrant investigation with more comprehensive experiments?*
2. *What are the likely economic benefits from the use of fertilizer and are they sufficient to justify the costs and bother that are involved in buying and applying it?*
3. *How reliable are the estimates of expected benefits obtained with the experiment?*
4. *What are the best rates of fertilizer application suggested by the experiment?*

To answer these questions requires the estimation of yield responses to the nutrient applications in the presence and absence of each other, as shown by the ΔY values in Table 2.5 calculated from estimates of yield for the treatment rates provided by the equation (2.6) and corresponding estimates of profit responses $\Delta \Pi$ for the increase in economic returns after allowance for all relevant economic factors, as given later in Table 2.7. Answers to these questions for the example experiment are given at the end of the chapter.

Table 2.5 Estimates of yield calculated from equation (2.6) and the differences, ΔY_n for response to N and ΔY_p for response to P.

kg N/ha	kg P/ha		ΔY_p
	0	50	
0	1320.3	2977.0	1656.7
100	1336.0	3420.0	2084.0
ΔY_n	15.7	443.0	

Economic variables

Perhaps the most unattractive feature of the above type of questions to the scientist is that to answer them requires the application of procedures, even though simple, which belong to another discipline, in this case of economics. Nevertheless farmers have to answer such questions as best possible every time they apply fertilizer so that it does not seem too much to expect scientists, with their education and technical training, to do the same. In any case the estimations require only simple calculations, using local estimates of fertilizer costs and crop value and a discounting to allow for the time interval between when money is invested in fertilizer application and when a return is received after the crop has been grown, harvested and sold.

Fertilizer costs

The most important factor affecting the use of fertilizers is their cost so that fertilizer application rates must be chosen on the basis of economic returns rather than simply for the correction of soil nutrient deficiencies. A first requirement then, in studies on the practical significance of the results of fertilizer experiments, is an estimation of the costs of fertilizing in the region represented by experiments and then an estimation of the economic returns from various rates of fertilizer application in order to estimate fertilizer requirements for maximum economic benefits.

The price of fertilizers at a local source is easily determined but gives only the first part of the cost of fertilizing. In addition there is the cost of transport of the fertilizer to the farm and this can be considerable and even prohibitive for farms located in remote regions.

Table 2.6 Cost[1] components of fertilizers and crop affecting economic return from fertilizing.

	Urea	Super-phosphate	Crop
Market price, $/t	301.2	276.7	98.0
Transport cost, $/t	26.0	26.0	10.0
Storage, handling, $/t	4.0	4.0	8.0
Price to farmer, $/t	331.2	306.7	80.0
Fertilizer composition	46.0%N	16.0%P	
Price per kg of nutrient or crop	C_n = $0.72	C_p = $1.92	V = $0.08

Then allowance must be made for the cost of labour in the handling and storage of the fertilizer which although typically small relative to the other costs should be included, and indicated, if only to reassure farmers that economic allowance has been made for the "bother" of using fertilizer. Thus for the example costs in Table 2.6, a cost of transport ($26/tonne)[1] and a cost for handling and storage ($4/tonne) have been added to the costs of fertilizer. For calculation purposes the sum of all such costs is conveniently expressed on the basis of cost per unit of nutrient in the fertilizer, as at the bottom of the table, to facilitate comparisons with alternative fertilizers.

Crop value

The value of a crop is estimated similarly with allowances for all costs. Thus for the example, a market price of $98 per tonne of crop yield is reduced by $18 + $8 per tonne to allow for harvest, transport and storage costs giving an economic value to the farmer of V = $0.08 per kg of harvested crop.

[1] For the purposes of the economic examples here and in the following chapters, prices and costs are shown with a $ sign, simply to indicate a unit of currency. Although based on Australian values and dollars at the time of writing, the values are chosen only to illustrate calculation procedures and not to suggest actual values or a particular currency.

Fixed costs

Farming involves many costs which are incurred whether fertilizer is applied or not, as for land preparation, cultivation, interest on mortgages, depreciation of farm machinery, pest control, seed, etc. Since these costs are independent of amount of fertilizer applied or the amount of crop product, they are called *fixed costs*. Fixed costs are important because in conjunction with the costs of fertilizing, they determine whether a crop is worth producing or not. Since however they do not vary with the amount of fertilizer that is applied, they do not affect the economic return from an investment in fertilizer and they are not required for calculations of economic returns from fertilizer applications or of optimal rates of application. Thus when fixed costs are relatively small it is possible to calculate economically optimal fertilizer rates that will produce a maximum profit and equivalently, when fixed costs are prohibitively high, it is possible to calculate economically optimal rates that will produce a minimum loss. In this latter situation where even with optimal application rates crop production results in a loss, fertilizer use would be nil of course, simply because there would be no cropping.

Interest rate and minimal rate of return

If a farmer borrows money to purchase and apply fertilizer, the loan must be repaid together with interest, as calculated by the usual compound interest formula

$$B = A(1+R)^t \quad (2.11)$$

where A is the amount borrowed, R is the interest rate, t is the number of interest conversion periods for the duration of the loan and B is the amount that must be repaid. Thus if fertilizer is applied at the rate N (kg nutrient/ha) and the initial cost of fertilizing is C_n per unit of N, the final cost per ha, including the interest repayment, will be $C_nN(1+R)^t$.

Equivalently if the farmer uses money from his savings to cover the cost of fertilizing, in order to determine profit a deduction must be made from the amount received with the sale of the crop corresponding to the return he would have received if he had used the money for some other purpose. In this case the value for R is determined by estimating the rate of return from the best alternative investment opportunity for the money spent on fertilizing whereas in the previous case it is simply determined as an interest rate by the lender. Examples of alternative investment opportunities are an interest bearing deposit in a bank, pest control and also in a broader sense, investments giving non-quantifiable rates of return such as for the purchase of household furniture, the education of children, insurance and so on. Values of R determined by alternative investment opportunities are often difficult to specify, particularly when there are many alternative investments and limited money resources. For example a value $R = 0.2$ or 20% might be chosen rather arbitrarily when local bank interest rates for a deposit are 10%, given the risks associated with crop production and the many alternative demands on a farmer's savings. In general values of R chosen for poor farmers are higher than for rich farmers because poor farmers have smaller money resources than the rich and consequently cannot afford the luxury of making investments that produce relatively low rates of return. For example the poor farmer must farm with simple implements because they produce a relatively high rate of return relative to their cost (high value of R for an alternative investment) whereas the rich farmer can farm with sophisticated implements producing higher yields and profits because he has the economic resources to afford the lower rate of return relative to their cost (low value of R for an alternative investment).

Time discounting

The interest formula $B = A(1+R)^t$ can be rearranged to

$$A = \frac{B}{(1+R)^t} \qquad (2.12)$$

to give the formula for calculating the present value A of an amount B that will be received in the future, after t conversion periods, with the interest rate R. The present value A is smaller than the future value B so that the adjustment provided by this formula is termed a *time discounting* of B. Time discounting is used in the calculation of economic returns from fertilizer applications when there are successive returns from effects of the applications on successive crops or from the successive harvests of perennial crops as will be described in chapter 3.

Profit function

Profits produced by the use of fertilizer are calculated from functions using the above economic variables. Thus for the example experiment the yield model (2.2)

$$Y = b_0 + b_1N + b_2P + b_3NP + b_4L_b + b_5Q_b$$

where

Y = crop yield

N = rate of application of N fertilizer (kg N/ha)

P = rate of application of P fertilizer (kg P/ha)

can be used with the economic variables

V = crop value ($/kg production)

C_n = cost of N fertilizer ($/kg N)

C_p = cost of P fertilizer ($/kg P)

R = interest rate or rate of return for an alternative investment

t = number of interest conversion periods

Q_t = total of fixed costs after t conversion periods

to derive the profit function

$$\Pi_t = VY - C_n N(1+R)^t - C_p P(1+R)^t - Q_t \tag{2.13}$$

Here Π_t is the profit per unit area obtained t interest conversion periods after the investment in fertilizing, VY is the money received from the sale of the crop, and $C_n N(1+R)^t$ and $C_p P(1+R)^t$ are the costs of N and P applications with interest adjustments by (2.11) .

An alternative profit function can be derived using the time discounting formula (2.12) to obtain Π_0 for profit and Q_0 for fixed costs at the time of the investment

$$\Pi_0 = \frac{VY}{(1+R)^t} - C_n N - C_p P - Q_0 \tag{2.14}$$

The functions are equivalent for the calculation of optimal rates of fertilizer application, that is the same rate gives the maximum value for both Π_t and Π_0.

The calculation of profits by these functions is complicated by the presence of the terms Q_t or Q_0 for fixed costs but this is not important, as already indicated, for the calculation of economic returns from the application of fertilizer and consequently of optimal rates. Thus if Y_0 is the yield produced without fertilizer then the profit without fertilizer after t interest conversion periods is simply

$$\Pi_{t0} = VY_0 - Q_t$$

Deducting this profit from that obtained with fertilizer gives the increase in profit due to the use of fertilizer

$$\Delta\Pi_t = \Pi_t - \Pi_{t0} = V(Y - Y_0) - C_n N(1+R)^t - C_p P(1+R)^t \tag{2.15}$$

with the fixed cost term disappearing with the subtraction. A corresponding function for $\Delta\Pi_0$, without a fixed cost term, can be derived from (2.14).

Calculation details for the use of these functions are illustrated in Table 2.7 using the regression estimates of Y from Table 2.5, the economic values C_n, C_p and V from Table 2.6 and an assumed value $R = 0.2$ for rate of return from the best alternative investment opportunity. No estimate is made of fixed costs so that profits without this deduction are shown in the columns $\Pi_t + Q_t$ and $\Pi_0 + Q_0$. Comparisons of the profits Π_t and Π_0, and the profit increases $\Delta\Pi_t$ and $\Delta\Pi_0$, all lead to the same conclusions concerning the economic benefits to be obtained from the use of fertilizer. All show a clear economic benefit from the application of P at the rate 50 kg N/ha but economic losses from the application of N at the rate 100 kg N/ha. Comparisons also show a small benefit from the small, non-significant, positive N × P interaction effect, the loss from the N application being less in the presence of the P application and the gain from the P application greater in the presence of the N application.

Table 2.7 Profit calculations by functions (2.6) to (2.8).

N	P	Y kg/ha	VY $/ha	C_nN $/ha	C_pP $/ha	$\Pi_t + Q_t$ $/ha	$\Delta\Pi_t$ $/ha	$\Pi_0 + Q_0$ $/ha	$\Delta\Pi_0$ $/ha
0	0	1320	105.6	0.0	0.0	105.6	0.0	88.0	0.0
0	50	2977	238.2	0.0	96.0	123.0	17.4	102.5	14.5
100	0	1336	106.9	72.0	0.0	20.5	−85.1	17.1	−70.9
100	50	3420	273.6	72.0	96.0	72.0	−33.6	60.0	−28.0

The economic values calculated in this way provide bases for answering the above practical questions about the results of the example experiment. For example:

1. *Are the yield responses to fertilizer applications large enough to warrant further investigation with more comprehensive experiments?*

There are statistically and economically significant yield responses to P that warrant further studies. Corresponding responses to N are small and would only warrant further consideration if there are reasons for expecting larger responses in other situations, as in other years, at other sites or with other treatment rates.

2. *What are the likely economic benefits from the use of fertilizer and are they sufficient to justify the costs and bother that are involved in buying and applying it?*

The experiment demonstrates a gain of $14.5/ha from an application of 50 kg P/ha after allowing for relevant costs including handling costs. Larger gains may be expected from other application rates but the present experiment gives no basis for estimating these rates or gains.

3. *How reliable are the estimates of expected benefits obtained with the experiment?*

The effect of the P fertilizer is statistically highly significant. The estimates for the P effects can consequently be considered to be very reliable for the conditions represented by the experiment.

4. *What are the best rates of fertilizer application indicated by the experiment?*

The experiment shows the highest gain from an application of 50 kg P/ha with no N fertilizer. Better estimates of the best or optimal rate would be obtained from experiments with more treatment rates.

Chapter 3
Optimal Fertilizer Application Rates

3.1 The optimal rate

Fertilizer applications increase crop yields by removing soil nutrient deficiencies but the basic reason that farmers apply fertilizers is to *increase economic returns* from cropping - the increases in yield are the means to this end. Accordingly we will adopt the simple definition:

> *The optimal rate of application of fertilizer to a crop is that rate which produces maximum economic return.*

Optimal rates determined on the economic basis of this definition are smaller than those required to produce maximum crop yields. Consequently, in addition to the economic benefits they produce, they are less likely to produce undesirable effects on the environment as, for example, by the pollution of ground waters with residues of fertilizer not used by crops.

The definition serves to indicate procedures for calculating optimal rates. Thus in general terms, given the simple case where the relationship between fertilizer application rate X and profit from cropping Π is represented by the functional relationship $\Pi = f(X)$, then the optimal rate for X is calculated by the usual differential calculus procedure of differentiating to obtain the differential $d\Pi/dX$, equating with zero, and then solving the equation $d\Pi/dX = 0$ for X, that is for the value of X that produces the maximum of Π. Similarly if the relationship between profit and rates of application of the several nutrients N, P, ... is represented by the multinutrient relationship $\Pi = f(N, P, \ldots, S)$, then optimal rates for $N, P, \ldots$ are calculated by solving equations $\partial\Pi/\partial N = \partial\Pi/\partial P = \ldots = \partial\Pi/\partial S = 0$. The main problem in calculating optimal rates is the estimation of profit functions $\Pi = f(X)$, $\Pi = f(N, P, \ldots)$, etc. to represent the profit fertilizer rate relationship for particular crops and cropping systems as described in following chapters rather than the calculation procedures that follow from the definition as described in this chapter.

3.2 General procedure

If a function $Y = f(X)$ has a maximum, then the value of X that produces this maximum is calculated by obtaining the derivative dY/dX and then solving the equation $dY/dX = 0$ for X, checking that the solution coincides with the maximum rather than a minimum or saddle point. For example given a fertilizer-yield function

$$Y = b_0 + b_1 N^{.5} + b_2 N \tag{3.1}$$

relating crop yield Y to the fertilizer nutrient application rate N, then $\dfrac{dY}{dN} = 0$ gives $\dfrac{b_1}{2N^{.5}} + b_2 = 0$ thus producing the formula

$$N = \left(\frac{-\,b_1}{2b_2} \right)^2 \tag{3.2}$$

for calculating the value N for maximum yield. For example if the fertilizer nutrient is phosphorus and the relationship is

$$Y = 600 + 500P^{.5} - 30P \tag{3.3}$$

then the formula (3.2) gives the value of P producing maximum Y

$$P = \left(\frac{-500}{(2)(-30)} \right)^2 = 69.4$$

corresponding to the phosphorus application rate for the maximum yield illustrated with the graph for this function in Fig. 3.1.

The same general procedure is followed to calculate the optimal nutrient application rate N that gives the maximum profit Π, from the profit function $\Pi = f(N)$. Thus given a single nutrient profit function

$$\Pi = VY - C_n N(1+R)^t - Q \tag{3.4}$$

corresponding to equation (2.13) of the previous chapter where V is value of a unit of crop yield Y, C_n is cost of a unit of fertilizer nutrient, R is interest rate or rate of return, t is number of interest conversion

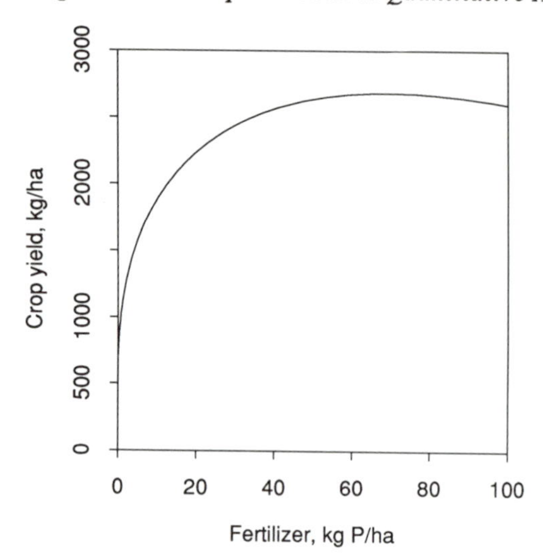

Fig. 3.1 Functional relationship $Y = 600 + 500P^{.5} - 30P$ between crop yield and application rate of phosphorus in fertilizer.

periods for time discounting and Q is fixed costs, the equation for optimal rate is derived from the equation defined by $\frac{d\Pi}{dN} = 0$. Since $\frac{d\Pi}{dN} = V\frac{dY}{dN} - C_n(1+R)^t$, this equation can be replaced by $V\frac{dY}{dN} - C_n(1+R)^t = 0$ to obtain the more convenient form

$$\frac{dY}{dN} = \frac{C_n}{V}(1+R)^t$$

The right hand side of this equation contains all of the economic variables required for the calculation of optimal rates so that the general economic variable E_n can be defined by

$$E_n = \frac{C_n(1+R)^t}{V} \tag{3.5}$$

to represent the economic situation that is relevant to the calculation of the optimal rate. The equation

$$\frac{dY}{dN} = E_n$$

thus provides the general basis for the calculation of optimal rates from yield functions $Y = f(N)$. It may be noted incidentally that the fixed costs term Q in (3.4) disappears with the differentiation and consequently does not appear in the expression for E_n corresponding with the fact noted in chapter 2 that fixed costs, although important economic variables, are not relevant to the calculation of economic rates. Also it will be found that the identical equation $dY/dN = E_n$ is derived from corresponding functions for *increase* in profit due to fertilizer. Thus if the increase in profit function

$$\Delta\Pi = V(Y - Y_0) - C_n(1+R)^t$$

is derived from the profit function (3.4) where Y_0 is yield without fertilizer, $d(\Delta\Pi)/dN = 0$ leads to the identical equation $dY/dN = E_n$ since Y_0 is not a function of N. The disappearance of Y_0 with the differentiation is noteworthy since it shows that there is no need to estimate the yield that is produced without fertilizer for the calculation of optimal rates from fertilizer-yield functions.

The convenience of the $dY/dN = E_n$ equation for the derivation of optimal rate equations can be shown with the yield function

$$Y = b_0 + b_1 N^{.5} + b_2 N$$

For this function

$$\frac{dY}{dN} = \frac{b_1}{2N^{.5}} + b_2$$

so that the optimal rate equation $dY/dN = E_n$ becomes

$$\frac{b_1}{2N^{.5}} + b_2 = E_n$$

and algebraic rearrangement gives the formula for the optimal rate

$$N = \left(\frac{0.5b_1}{E_n - b_2}\right)^2 \qquad (3.6)$$

For example given the example yield function

$$Y = 600 + 500P^{.5} - 30P$$

where the nutrient is phosphorus and the example economic values are $V = 0.08$, $C_p = \$1.92$, $R = 0.2$ and $t = 1$, the profit function corresponding to (3.4) is

$$\Pi = VY - C_p P(1+R)^t - Q$$

giving by substitution and algebraic rearrangement

$$\begin{aligned} \Pi &= (0.08)(600 + 500P^{.5} - 30P) - 1.92P(1 + 0.2)^1 - Q \\ &= 48 + 40P^{.5} - 4.7P - Q \end{aligned} \qquad (3.7)$$

The example values give the value for the general economic variable

$$E_p = \frac{C_p}{V}(1+R)^t = \frac{(1.92)(1+0.2)^1}{0.08} = 28.8$$

Applying the optimal rate formula (3.6), the optimal rate producing the maximum value for Π is calculated by

$$P = \left(\frac{0.5b_1}{E_p - b_2}\right)^2 = \left(\frac{(0.5)(500)}{28.8 - (-30)}\right)^2 = 18.1 \text{ kg P/ha}$$

corresponding to the application rate for maximum profit in Fig. 3.2.

Alternatively the profit increase equation can be derived,

$$\Delta\Pi = 40P^{.5} - 4.7P \qquad (3.8)$$

leading to the same optimal rate, 18.1 kg P/ha.

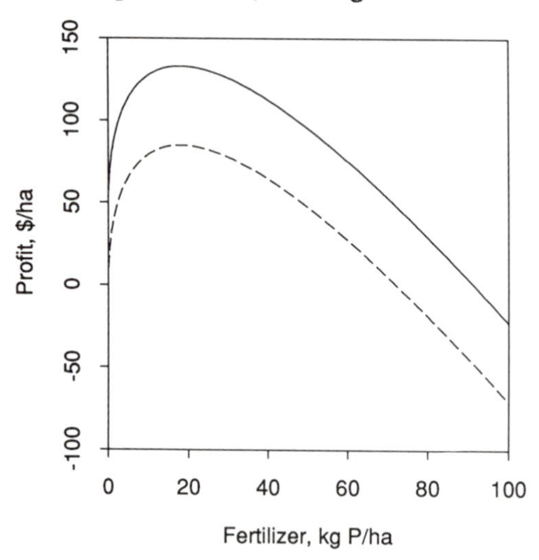

Fig. 3.2 Example profit and profit increase functions (3.7) and (3.8), solid and dashed lines respectively, derived from the yield function of Fig. 3.1. The graph for the profit function (3.7) is without adjustment (downwards) for the fixed costs Q.

The derivation of the optimal rate equation by $dY/dN = E_n$ for the function $Y = b_0 + b_1 N^{.5} + b_2 N$ has been detailed because this function provides a convenient mathematical basis for the derivation of general soil fertility models as described in later chapters. It should be understood however that the procedure is general and may be applied to other functional forms for the yield-fertilizer rate relationship, as illustrated in Table 3.1 with some alternative forms for $Y = f(N)$. With these examples the quadratic functions 1 and 2, the classical Mitscherlich function 3 and its extension as the "exponential plus linear" function 4 give simple and direct calculation formulas. Values calculated for the quadratic function 2 will generally be misleading however, because this particular function is not suitable for the representation of the relationship as will be described in chapter 5. The function 5 is mathematically interesting in that it gives only two possible optimal rates, b_2 or zero, but it also is judged to be unsuitable

Table 3.1 Formulas for the calculation of optimal fertilizer rates for alternative forms for the functional relationship $Y = f(N)$.

	Functional form $Y = f(N)$	Optimal rate by $dY/dN = E_n$
1.	$Y = b_0 + b_1 N^{.5} + b_2 N$	$N = \left(\dfrac{0.5 b_1}{E_n - b_2}\right)^2$
2.	$Y = b_0 + b_1 N + b_2 N^2$	$N = \dfrac{E_n - b_1}{2 b_2}$
3.	$Y = b_0 + b_1 e^{b_2 N}$	$N = \dfrac{1}{b_2} \log\left(\dfrac{E_n}{b_1 b_2}\right)$
4.	$Y = b_0 + b_1 e^{b_2 N} + b_3 N$	$N = \dfrac{1}{b_2} \log\left(\dfrac{E_n - b_3}{b_1 b_2}\right)$
5.	$Y = b_0 + b_1 (N - b_2 - \lvert N - b_2 \rvert)$	$N = b_2$, if $2b_1 \geq E_n$ $N = 0$, if $2\text{b}_1 < E_n$
6.	$\log Y = b_0 + b_1 e^{b_2 N}$	$b_1 b_2 \exp\{b_0 + b_2 N + b_1 \exp(b_2 N)\} = E_n$

for the representation of the relationship and hence for the calculation of optimal rates. The function 6 is included as an example of many alternative functions which although giving good representations of the relationship, are not likely to be popular because $dY/dN = E_n$ produces difficult optimal rate equations that must be solved by successive approximation procedures.

The derivation of equations for calculating optimal fertilizer rates for more complex situations follows this same type of procedure, as illustrated in following sections. First however the present example is used to illustrate features of yield and profit functions relating to optimal fertilizer rates.

3.3 Features of yield and profit functions

Characteristic features associated with the calculation of optimal rates are illustrated in Table 3.2 with columns of values for yield and economic variables calculated from the example yield function (3.3) and the corresponding profit function (3.7), for the range of nutrient application rates $P = 10$ to 22 kg P/ha in the vicinity of the optimal rate, $P = 18.1$ kg P/ha. Values are also given for the rate $P = 69.4$ that produces the maximum yield. Columns are for:

P = application rate of the fertilizer nutrient,

Y = yield calculated from the yield function $Y = 600 + 500P^{.5} - 30P$,

$\Pi + Q$ = profit without deduction for fixed costs (solid line in Fig. 3.2), calculated from $\Pi = 48 + 40P^{.5} - 4.7P - Q$,

$\Delta\Pi$ = gain due to fertilizer (broken line graph in Fig. 3.2), calculated from $\Delta\Pi = 40P^{.5} - 4.7P$,

$Cost = C_pP(1+R)$ for cost of fertilizing,

$d\Pi/dP$ = rate of economic gain due to P,

dY/dP = rate of increase in yield with increase in P, and

$\Delta\Pi/Cost$ = "Benefit/Cost" ratio.

Diminishing response form

Profit functions must define a maximum if they are to be suitable for the calculation of optimal rates and it follows that profit and yield functions must have a diminishing response form such that Π and Y

Table 3.2 Values calculated from the example equations (3.3), (3.7) and (3.8) to illustrate features associated with the optimal rate.

P kg/ha	Y kg/ha	$\Pi + Q$ $/ha	$\Delta\Pi$ $/ha	$Cost$ $/ha	$d\Pi/dP$	dY/dP	$\Delta\Pi/Cost$
10	1881	127.45	79.45	23.04	1.62	49.1	3.45
12	1972	130.12	82.12	27.65	1.07	42.2	2.97
14	2051	131.81	83.81	32.26	0.64	36.8	2.60
16	2120	132.74	84.74	36.86	0.30	32.5	2.30
18	2181	133.03	85.03	41.47	0.01	28.9	2.05
18.1	2184	133.03	85.03	41.70	0.00	28.8	2.04
20	2236	132.81	84.81	46.08	–0.23	25.1	1.84
22	2285	132.12	84.12	50.69	–0.44	23.3	1.66
69.4	2683	54.67	6.67	160.00	–2.30	0.0	0.04

have maximum values and the rates of response $d\Pi/dN$ and dY/dN decrease progressively towards zero with increase in fertilizer rate. Thus for the example function, $d\Pi/dP$ decreases to $d\Pi/dP = 0$ with increase in P to $P = 18.1$ for the optimal rate as defined above and dY/dP decreases to $dY/dP = 0$ with increase to $P = 69.4$ for the rate producing maximum yield. The example profit and yield functions have a diminishing response form as illustrated in Figs 3.1 and 3.2 and as shown by the successively smaller increments of yield and of profit ($\Delta\Pi$) to the successive increments of 2 kg P/ha with the application rates in Table 3.2. Functions with these features are described as obeying the *Law of Diminishing Returns.*

Benefit/Cost ratio

The diminishing response form of profit functions has an important significance for the popular "Benefit/Cost" ratio of economics in that this ratio is both *misleading* and *inappropriate* if applied in relation to benefits gained from the use of fertilizers. This is shown by comparing the benefit/cost ratio as represented by the ratios in the $\Delta\Pi/Cost$ column in Table 3.2 with the profits in the $\Pi + Q$ and $\Delta\Pi$ columns. Any application of fertilizer at a rate greater than the optimal rate, $P = 18.1$, produces lower profits as would be expected, whereas the $\Delta\Pi/Cost$ ratios mislead by suggesting continuing and considerable economic benefits from higher rates. For example with an application of 22 kg P/ha, the "Benefit/Cost" ratio $\Delta\Pi/Cost$ still has the considerable value of 1.66, suggesting a clear benefit from applying fertilizer at this rate whereas in fact the "Benefit" is \$84.12/ha, that is \$0.91/ha less than the \$85.03/ha that is obtained if fertilizer is applied at the optimal rate of 18.1 kg P/ha. The Benefit/Cost ratio must not be used as a guide to optimal rates when there is a diminishing response type of relationship because it may easily lead to the overestimation of economic fertilizer rates as in this example. Apart from the economic considerations the overestimation of fertilizer requirements due to the use of this ratio may produce undesirable environmental effects from fertilizer residues.

Economic elasticity

Profits obtained with a range of fertilizer application rates in the vicinity of the optimum often show a puzzling feature due to the diminishing response form of the profit functions. Profit increments become smaller as application rates approach the optimum so that profits become nearly constant for a wide range of fertilizer rates and, correspondingly, for a wide range of investments in fertilizing. For example the values in Table 3.2 show a maximum increase in profit $\Delta\Pi$= \$85.03/ha from the optimal application rate of 18.1 kg P/ha with *Cost* = \$41.70/ha but this profit is little different from $\Delta\Pi$= \$84.74/ha produced by 16 kg P/ha with *Cost* = \$36.86/ha or the \$84.12/ha produced by 22 kg P/ha with *Cost* = \$50.69. Such a relative constancy of economic returns (\$84.12 to \$85.03/ha) over a wide range of investments (\$36.86 to \$50.69/ha) is described as *high economic elasticity*. In general, elasticity = $(X/Y)(dY/dX)$ for a function $Y = f(X)$ (Tintner and Millham 1969).

The degree of economic elasticity associated with the use of fertilizer depends on the economic situation and the nature of the functional relationship between yield and fertilizer rate, elasticity being high when the slope of a yield function dY/dN is close to the value of the economic variable E_n (3.5) for a wide range of application rates. This is illustrated in Figs 3.3 and 3.4 with graphs for the yield function

$$Y = 300 + 40P^{.5} + 26P$$

and the corresponding profit increase function

$$\Delta\Pi = 3.2P^{.5} - 0.224P$$

derived with the example values $V = 0.08$, $C_p = 1.92$ and $R = 0.2$. With this function the slope dY/dP is close to the economic variable value $E_p = 28.8$ over a wide range of values for the nutrient application rate P. Thus for the wide range of application rates from 20 to 100 kg P/ha,

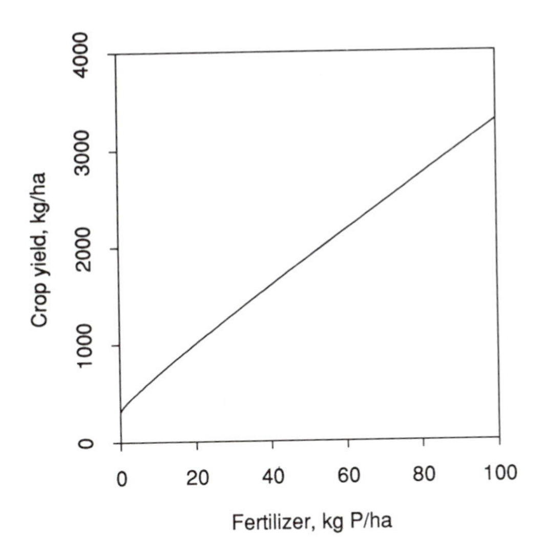

Fig. 3.3 A yield function with slope *dY/dP* close to the value of the economic factor E_p over a wide range of values for *P*.

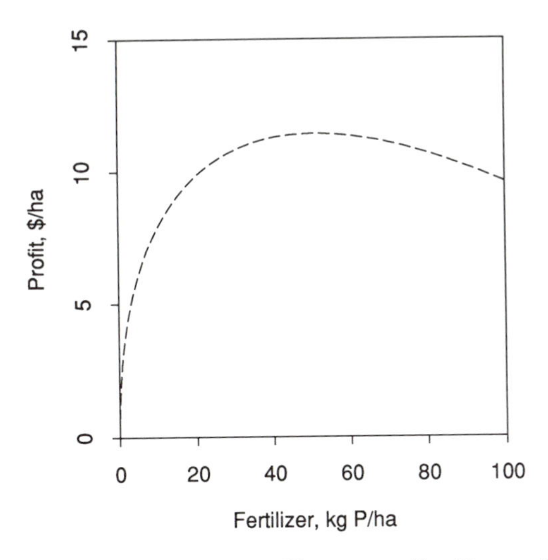

Fig. 3.4 Profit from fertilizer applications for the yield function of Fig. 3.3, showing high economic elasticity in the vicinity of the optimal rate P = 51.

the rate of response dY/dP ranges only from 30.5 to 28.0 and $\Delta\Pi$ ranges from \$9.83 and \$9.60 for $P = 20$ and 100 respectively, to \$11.43/ha for the optimal rate of 51.02 kg P/ha whereas for this same range the cost of fertilizing ranges widely from \$38.40 for $P = 20$ to \$192.00/ha for $P = 100$. In this contrived example there is thus a very high elasticity, economic gains ranging only from \$9.83 to \$11.43/ha for the very wide investment range, \$38.40 to \$192.00/ha.

The important point to remember concerning the typically high elasticity of fertilizer-profit functions is that the value of the variable C_n for fertilizer cost in profit equations includes all costs relating to the application of fertilizer so that although there may be little economic gain from using the optimal rate rather than a somewhat lower rate, there is also no economic advantage in choosing a lower rate. If the yield function is a statistical best estimate then the optimal rate calculated from this estimate is also a best estimate. Given the inevitable error associated with any such estimate whereby it may be either too high or too low, it is only sensible to use the best estimate since it can be expected to be near the centre of the probability range for the actual optimal rate.

Sensitivity to economic variables

Optimal rates are affected by variations in the values for the economic variables V, C_n, R and t and the relative magnitude of these effects, called the *sensitivity* of optimal rates to the economic variables, can be

Table 3.3 Effects of variations of values for the economic variables in the economic factor E_p on the optimal rate calculated by the formula (3.6) with the example yield function (3.3).

C_p	V	R	t	E_p	Optimal P
1.92	0.08	0.2	1	28.8	18.1
3.84	0.08	0.2	1	57.6	8.1
0.96	0.08	0.2	1	14.4	31.7
1.92	0.16	0.2	1	14.4	31.7
1.92	0.04	0.2	1	57.6	8.1
1.92	0.08	0.1	1	26.4	19.6
1.92	0.08	0.4	1	33.6	15.5
1.92	0.08	0.2	0.5	26.3	19.7
1.92	0.08	0.2	2	34.6	15.0

investigated with calculations with alternative values for the variables as illustrated in Table 3.3 with optimal P values calculated for the example yield function (3.3). To show sensitivities, the value of each of the variables has been doubled and halved, each in turn, to show the consequent effects on optimal rates. Comparisons of the optimal rates in this way show that they are more sensitive to variations in V for crop value and C_p for cost of the nutrient P than to R for interest rate or t for time conversion periods.

3.4 Simultaneous optimal rates for multiple nutrient deficiencies

When there are several soil nutrient deficiencies, the deficiencies may be corrected simultaneously by applying each of the nutrients simultaneously in a mixture of appropriate single nutrient fertilizers. The set of optimal rates in the ideal mix, producing the maximum profit with the simultaneous application, is termed the *simultaneous optimal rates* for the multiple nutrient deficiencies. Procedures for calculating these simultaneous rates are obtained by a direct extension of that used for the single nutrient optimal rate, that is by using multiple nutrient functions $Y = f(N, P, \cdots, S)$ or $\Pi = f(N, P, \cdots, S)$ in place of the single nutrient functions $Y = f(N)$ or $\Pi = f(N)$ and simultaneous equations of partial differentials for the several nutrients, $\partial\Pi/\partial N = \partial\Pi/\partial P = \ldots \; \partial\Pi/\partial S = 0$ in place of the single nutrient equation $d\Pi/dN = 0$. Again the economic variables in these equations may be combined in single economic variables for each nutrient, E_n, E_p, ... , E_s so that these equations may be conveniently replaced by $\partial Y/\partial N = E_n$, $\partial Y/\partial P = E_p$, ... , $\partial Y/\partial S = E_s$.

If the effects of the different nutrients on crop production are independent of each other, the optimal rate for each nutrient can be determined individually from simple $\partial Y/\partial N = E_n$ type equations corresponding to the single nutrient equation $dY/dN = E_n$. Since however simultaneous applications of different nutrients usually produce interaction effects such that effects of the application of each nutrient are affected by applications of others, solutions to the equations must be obtained simultaneously. The calculation formulas for the simultaneous optimal rates become increasingly complex with increase in the number of nutrients as illustrated with the following examples for square root quadratic yield functions for one to three nutrients.

One nutrient

Model: $Y = b_0 + b_1 N^{\cdot 5} + b_2 N$

Optimal N by $\dfrac{dY}{dN} = E_n$

$$N = \left(\frac{0 \cdot 5 b_1}{E_n - b_2} \right)^2 \tag{3.9}$$

$$\text{where } E_n = \frac{C_n (1+R)^t}{V}.$$

Two nutrients

Model: $Y = b_0 + b_1 N^{\cdot 5} + b_2 P^{\cdot 5} + b_3 (NP)^{\cdot 5} + b_4 N + b_5 P$

Optimal N and P by $\dfrac{\partial Y}{\partial N} = E_n, \ \dfrac{\partial Y}{\partial P} = E_p$

$$\left. \begin{aligned} N &= \left(\frac{2b_1 (E_p - b_5) + b_2 b_3}{D} \right)^2 \\ P &= \left(\frac{2b_2 (E_n - b_4) + b_1 b_3}{D} \right)^2 \end{aligned} \right\} \tag{3.10}$$

$$\text{where } D = 4(E_n - b_4)(E_p - b_5) - b_3^2$$

$$\text{and } E_n = \frac{C_n (1+R)^t}{V}, \ E_p = \frac{C_p (1+R)^t}{V}$$

Three nutrients

Model: $$Y = b_0 + b_1 N^{.5} + b_2 P^{.5} + b_3 K^{.5} + b_4 (NP)^{.5} + b_5 (NK)^{.5} + b_6 (PK)^{.5} + b_7 N + b_8 P + b_9 K$$

Optimal N, P and K by $\dfrac{\partial Y}{\partial N} = E_n$, $\dfrac{\partial Y}{\partial P} = E_p$, $\dfrac{\partial Y}{\partial K} = E_k$

$$\left.\begin{aligned} N &= \left(\frac{b_1 b_6^2 - b_2 b_5 b_6 - b_3 b_4 b_6 - 2b_3 b_5 (E_p - b_8) - 2b_2 b_4 (E_k - b_9) - 4b_1 (E_p - b_8)(E_k - b_9)}{D} \right)^2 \\ P &= \left(\frac{b_2 b_5^2 - b_1 b_5 b_5 - b_3 b_4 b_5 - 2b_3 b_6 (E_n - b_7) - 2b_1 b_4 (E_k - b_9) - 4b_2 (E_n - b_7)(E_k - b_9)}{D} \right)^2 \\ K &= \left(\frac{b_3 b_4^2 - b_1 b_4 b_6 - b_2 b_4 b_5 - 2b_2 b_6 (E_n - b_7) - 2b_1 b_5 (E_p - b_8) - 4b_3 (E_n - b_7)(E_p - b_8)}{D} \right)^2 \end{aligned}\right\} \quad \ldots (3.11)$$

where

$$D = 2b_4 b_5 b_6 + 2b_6^2 (E_n - b_7) + 2b_5^2 (E_p - b_8) + 2b_4^2 (E_k - b_9) - 8(E_n - b_7)(E_p - b_8)(E_k - b_9)$$

and

$$E_n = \frac{C_n (1+R)^t}{V}, \quad E_p = \frac{C_p (1+R)^t}{V}, \quad E_k = \frac{C_k (1+R)^t}{V}$$

Four or more nutrients

The derivation of formulas for optimal rates can be extended mathematically to four or more simultaneous rates, given appropriate yield functions. Such formulas are rarely likely to be required in practice however because multinutrient yield functions that have been estimated from experimental data usually do not have the necessary diminishing form with respect to more than about two or three nutrients for the valid application of the calculation procedure. The types of problem preventing valid calculations are described in the following section. Even with three nutrient functions such problems are common so that in practice the equations (3.11) can seldom be used[1].

[1] Based on experiences with many four nutrient experiments in Australia (Colwell 1977, 1979) and Brazil.

3.5 Calculation hazards

Although the above mathematical procedures for calculating optimal nutrient application rates produce mathematically valid values, they often do not produce valid estimates of optimal fertilizer rates. Checks are therefore required to ensure that calculated values correspond to valid estimates of optimal rates[2]. To illustrate the hazards that arise with the undiscerning use of calculation formulas, and the associated traps for the unwary, the following very different fertilizer-yield functions have been devised, each to give the identical calculated value P = 72 with the optimal rate formula (3.9) and the economic variable value E_p = 28.8.

$$Y = 1600 + 190P^{\cdot 5} + 40P \quad \ldots \text{A}$$

$$Y = 1600 - 190P^{\cdot 5} + 40P \quad \ldots \text{B}$$

$$Y = 600 - 190P^{\cdot 5} + 17{\cdot}6P \quad \ldots \text{C}$$

$$Y = 600 - 190P^{\cdot 5} + 17{\cdot}6P \quad \ldots \text{D}$$

These functions represent very different types of relationship, as shown by the graphs in Fig. 3.5, but they nevertheless correspond with the types of function that are often estimated from the data of fertilizer experiments, especially when there has been a bad choice of treatment rates or when experimental conditions have been abnormal. Thus functions A and C represent the types of function that may be estimated when fertilizer effects are much greater than anticipated so that the range of fertilizer rates is not sufficient to define a maximum other than by extrapolation. Similarly, non-diminishing functions of the type B and D may be estimated when data, as well as not defining a clear maximum, are also affected by chance error effects, type B when the range of treatments has been too small to encompass the maximum and type D when there has simply been little or no response to the fertilizer treatments. For these example functions the variable value P = 72 is a valid estimate of an optimal nutrient application rate only for the function C although in each case this value is calculated from the optimal rate equation derived from $d\Pi/dP = 0$ or $dY/dP = E_p$. This is not obvious from a simple inspection of the functions and checks as described below are necessary to identify the invalid estimates. In general corresponding checks are necessary to identify functions for which the calculated values provide valid estimates of the optimal rate,

[2] For computer programs in Fortran that compute optimal rates with the formulas (3.9) to (3.11), with checks for validity, see Colwell (1978).

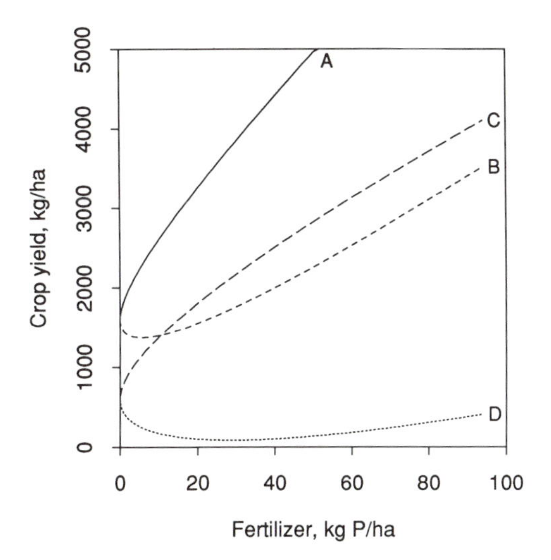

Fig. 3.5 Graphs of the above yield functions all of which give the value P = 72 by the optimal rate equation (3.9).

especially for the simultaneous optimal rates that are calculated from multinutrient functions.

Calculation checks

An inspection of the graphs of estimated yield functions may indicate when optimal rate formulas will give valid or invalid estimates of optimal rates but the interpretation of graphs for this purpose is often not simple, especially with multinutrient functions. Accordingly the following mathematical checks should be made when calculating optimal rates by the above formulas for square root fertilizer-yield functions and corresponding checks should be made with calculations for other types of function. With multinutrient functions the checks should be made for each nutrient in turn since if there are interaction effects, an invalid estimate for any one nutrient will invalidate the calculations for all of the others.

Check 1: Maximum or minimum Solutions of equations of the form $dY/dX = 0$ are used to calculate values of X that give a maximum, minimum or saddle point of a function $Y = f(X)$ and it is necessary to check which of these alternatives corresponds to calculated values. This can be done by simply comparing values for Y with values of X on either side of the calculated value of X or in the case of fertilizer-profit functions $\Pi = f(N)$ by calculating values for Π with values for the nutrient rate N on either side of the value calculated by the optimal rate formula. Alternatively the value for the second derivative may be calculated, a maximum being indicated if $d^2Y/dX^2 < 0$ and a minimum if $d^2Y/dX^2 > 0$ but this procedure will not indicate saddle points with certain types of function. For functions derived from the square root quadratic $Y = b_0 + b_1N^{.5} + b_2N$ the second derivative d^2Y/dN^2 indicates that the value calculated by the formula (3.9) is for a maximum if $b_1 > 0$ and for a minimum if $b_1 < 0$. Thus the example functions B and D are shown by this check not to be suitable for the calculation of optimal rates because the $P^{.5}$ coefficient in each is negative. Corresponding checks for values obtained for multinutrient functions by $\partial^2Y/\partial X^2$ type derivatives are more difficult for multinutrient functions because of the presence of interaction terms and because they may not identify a saddle point. In general then maxima and minima are best identified by the less sophisticated but surer procedure of comparing calculated profits for nutrient rates, above, below and at the calculated value for each of the nutrient rates.

Check 2: Positive or negative square root A problem arises with calculations from square root quadratic functions because the solution of the optimal rate equations involves a squaring with the consequence that the calculated value is always positive. This produces an ambiguity because the value that is squared may be either positive or negative, as for example with $(+2)^2 = (-2)^2 = +4$. For this reason the functions $Y = b_0 + b_1N^{.5} + b_2N$ and $Y = b_0 - b_1N^{.5} + b_2N$ produce identical solutions to the equation $dY/dP = E_p$. Problems with this ambiguity are avoided when calculating optimal rates by the formulas (3.9) to (3.11) by recognizing that regression estimates of the respective yield functions are made with only positive values for the square roots of the regressor variables values for nutrient rates. This calculation trap is avoided then by checking that bracketed expressions in the equations (3.9) to (3.11) are positive before they are squared when calculating optimal rates. If negative they simply correspond to values for mathematical extrapolations of the functions beyond the range of the data from which they were estimated. Thus for the function A, the expression

$$\left(\frac{0.5b_1}{E_n - b_2}\right) = \frac{(0.5)(190)}{(28.8 - 40)} = -8.48$$

in (3.9) is negative so that the value $P = (-8.48)^2 = 72$, being the square of a negative quantity, is not a valid estimate of the optimal rate.

Thus checks 1 and 2 show that of the example functions, only function C can be used to calculate the optimal rate with the optimal rate formula. Other checks are also necessary however before calculated values are accepted as estimates of optimal rates.

Check 3: Extrapolation error Fertilizer-yield functions are estimated from data covering particular ranges of fertilizer application rates and although these functions can be used to calculate estimates of yield, profit and optimal rates outside these ranges, such estimates are subject to error, called *extrapolation error*, the magnitude of which usually increases rapidly with the degree of extrapolation, especially with polynomial functions. Accordingly estimates should be checked for extrapolation. If the extrapolation is small, the calculated values may be accepted, cautiously, but otherwise they are best rejected as being too unreliable to indicate any more than that the optimal rate is above or below the range of the data as the case may be.

When simultaneous rates are calculated from multinutrient yield functions, all of the calculated rates are subject to extrapolation error if any calculated rate involves extrapolation because the effects of the different nutrients are not independent but connected by interaction effects. If for example simultaneous optimal rates are calculated for N, P and K by the equations (3.11), then all values are affected if any one of the calculated values involve an extrapolation, even if the calculated values for the others are well within the data range.

Check 4: Interpolation errors Most estimates made with fertilizer-yield functions are made by interpolation, that is within the range of the data from which the functions were estimated. These estimates may be affected by a much less obvious although common and sometimes serious error called *interpolation error*. For example if the function (3.3), illustrated in Fig. 3.1, had been estimated by the regression procedure using experimental data with treatment rates 0, 20, 50 and 100 kg P/ha, estimates of yield for rates between these treatment rates are by interpolation and their accuracy depends on the extent to which the estimated function represents the true relationship within this range. Serious interpolation errors can be produced when an inappropriate mathematical model has been chosen for functions to represent the relationship. In such cases regression estimates of yield may be good at fertilizer rates close to the treatment rates but misleading for other rates. Less obviously, estimates of optimal rates may be seriously affected by the use of an inappropriate model, even if the estimated rate is close to a treatment rate, because the slope dY/dN

of an estimated function for any value of N is affected by the choice of model as well as the data. The recognition of possible serious interpolation error due to the choice of model depends on knowledge about mathematical features that are imposed by the model, as described in chapter 4.

Estimates from non-ideal yield functions

Fertilizer experiments are carried out with the expectation that they will provide data that can be used to calculate estimates of fertilizer requirements without any of the types of problem with fertilizer-yield functions described above. In practice however, it is not uncommon for such problems to occur so that functions estimated from experimental data cannot be used to calculate optimal rates. When this happens the obvious solution to the problem is to simply reject the experiments that produce the wrong types of function and to repeat them with appropriate precautions and adjustments, in the hope of producing satisfactory functions next time. However such a solution is rather facile and ignores the practical problem that the repetition of fertilizer experiments is expensive and time consuming, involving a delay of at least a year to obtain a new and not necessarily better result. Consequently when functions do not have an ideal form it is still necessary to obtain at least some sort of an estimate which although not ideal, will at least give some indication of fertilizer requirements, if only to provide a basis for the choice of treatments for repeat experiments. Although optimal rates cannot be calculated directly from non-ideal functions, useful though cruder estimates are usually still possible with allowance for the nature of the problem. For example given functions of the type represented by the curves A and C in Fig. 3.5, an inspection of their graphs suggests that problems have arisen because treatment rates did not go high enough to define the optimal rate and that the optimal rate is therefore somewhat higher than the highest treatment rate of 100 kg P/ha. Similarly an inspection of graphs of the type represented by curve D indicates that the fertilizer requirement is either nil or very small. Such information is better than none, especially with multinutrient yield functions experiments where a non-ideal response form for one nutrient prevents the calculation of simultaneous optimal rates and hence the estimation of optimal rates for the other nutrients. In such cases reasonably good estimates can often be calculated for the other nutrient rates by substituting a conservative estimate for the nutrient producing the non ideal feature in the yield function and then calculating optimal rates for the others in the presence of this substituted rate. For example if an estimate of the three nutrient yield function

$$Y = b_0 + b_1 N^{.5} + b_2 P^{.5} + b_3 K^{.5} + b_4 (NP)^{.5} + b_5 (NK)^{.5} + b_6 (PK)^{.5}$$
$$+ b_7 N + b_8 P + b_9 K$$

cannot be used to calculate simultaneous optimal rates by formulas (3.11) because the yield response to P has the form B in Fig. 3.5, suggesting that the optimal rate is beyond a maximum treatment rate of 100 kg P/ha, a conservative estimate of the optimal rates for N and K can be obtained by substituting P = 100 in the estimated function. Thus substituting P = 100 in the above function gives

$$Y = b_0 + b_1 N^{.5} + b_2 (10) + b_3 K^{.5} + b_4 N^{.5} (10) + b_5 (NK)^{.5} + b_6 (10) K^{.5}$$
$$+ b_7 N + b_8 (100) + b_9 K$$

and hence

$$Y = (b_0 + 10 b_2 + 100 b_8) + (b_1 + 10 b_4) N^{.5} + (b_3 + 10 b_6) K^{.5} + b_5 (NK)^{.5}$$
$$+ b_7 N + b_9 K$$

This has the form of a two nutrient function

$$Y = c_0 + c_1 N^{.5} + c_2 K^{.5} + c_3 (NK)^{.5} + c_4 N + c_5 K$$

which can be used to calculate estimates of optimal rates for N and K, in the presence of P = 100, with the formula (3.10). Similarly if the yield response to P has the form D in Fig. 3.5, suggesting that the optimal rate is about zero, substituting P = 0 gives

$$Y = b_0 + b_1 N^{.5} + b_3 K^{.5} + b_5 (NK)^{.5} + b_7 N + b_9 K$$

also with the form of a two nutrient function that can be used to calculate optimal rates for N and K, this time in the absence of a P application. If interaction effects are not large, optimal rates for the other nutrients obtained in these ways can be expected to be reasonably accurate.

3.6 Optimal rates for multinutrient fertilizers

When there are several soil nutrient deficiencies farmers often wish to use a single fertilizer containing the several nutrients rather than to apply several single nutrient fertilizers at simultaneous optimal rates as calculated by the formulas of section 3.4. If mixed fertilizer can be purchased from a fertilizer blending plant that produces specified mixtures of single nutrient fertilizers, the simultaneous optimal rates can be used to indicate the ideal mix and the optimal rate of application of the mixed fertilizer. Alternatively farmers may have to purchase a multinutrient fertilizer with a predetermined composition and consequently have to choose a suitable fertilizer from several alternatives, each with different nutrient contents and with a different price. If fertilizer-yield functions

$$Y = b_0 + c_1 X^{\cdot 5} + c_2 X \tag{3.12}$$

can be estimated for each fertilizer then the optimal *fertilizer* rate can be calculated for each by the formula

$$X = \left(\frac{0.5c_1}{E_x - c_2} \right)^2$$

corresponding to that for the single nutrient rate formula (3.9), substituting X for N and fertilizer cost C_x for nutrient cost C_n to calculate the economic variable E_x. Profit with the optimal rate can then be calculated with the function for each fertilizer and the fertilizer and rate chosen that give the greatest profit. Although such functions can be estimated by experiments with treatments for each of the alternative fertilizers, it is more practicable when there are several alternatives to calculate functions for each fertilizer from multinutrient yield functions that have been estimated with multinutrient experiments. Thus for example given the three nutrient yield function

$$Y = b_0 + b_1 N^{\cdot 5} + b_2 P^{\cdot 5} + b_3 K^{\cdot 5} + b_4 (NP)^{\cdot 5} + b_5 (NK)^{\cdot 5} + b_6 (PK)^{\cdot 5}$$
$$+ b_7 N + b_8 P + b_9 K$$

the fertilizer function $Y = b_0 + c_1 X^{\cdot 5} + c_2 X$ for a multinutrient fertilizer containing N, P and K can be derived using analyses for its content of these nutrients.

The procedure for using the nutrient composition of a fertilizer to derive fertilizer-yield functions from multinutrient-yield functions is simply an extension of that used to calculate fertilizer rates from nutrient rates on the basis of their nutrient composition. Thus given an analysis a_n for a single nutrient fertilizer, equivalent rates are calculated by $X = N / a_n$ or $N = a_n X$ where X and N are respective rates of application of the fertilizer and the nutrient. For example if the fertilizer urea contains 46%N then 1 kg urea/ha is equivalent to 0.46 kg N/ha or 1 kg N/ha is equivalent to 2.17 kg urea/ha. The following conversion equations for deriving fertilizer functions from nutrient functions are derived by such $N = a_n X$ type substitutions.

One nutrient fertilizers

Given the one nutrient function

$$Y = b_0 + b_1 N^{.5} + b_2 N$$

and the fertilizer nutrient analysis a_n, the corresponding fertilizer-yield function is derived by substituting $a_n X = N$ to obtain

$$Y = b_0 + b_1 (a_n X)^{.5} + b_2 a_n X$$

or

$$Y = b_0 + c_1 X^{.5} + c_2 X \qquad (3.13)$$

where $c_1 = b_1 a_n^{.5}$ and $c_2 = b_2 a_2$

Two nutrient fertilizers

Given

$$Y = b_0 + b_1 N^{.5} + b_2 P^{.5} + b_3 (NP)^{.5} + b_4 N + b_5 P$$

and the analyses a_n and a_p, then

$$Y = b_0 + c_1 X^{.5} + c_2 X \qquad (3.14)$$

where $c_1 = b_1 a_n^{.5} + b_2 a_p^{.5}$ and $c_2 = b_3 a_n^{.5} a_p^{.5} + b_4 a_n + b_5 a_p$

Three nutrient fertilizers

Given

$$Y = b_0 + b_1 N^{.5} + b_2 P^{.5} + b_3 K^{.5} + b_4 (NP)^{.5} + b_5 (NK)^{.5} + b_6 (PK)^{.5}$$
$$+ b_7 N + b_8 P + b_9 K$$

and the analyses a_n, a_p and a_k then

$$Y = b_0 + c_1 X^{.5} + c_2 X \qquad (3.15)$$

where $c_1 = b_1 a_n^{.5} + b_2 a_p^{.5} + b_3 a_k^{.5}$

and $c_2 = b_4 a_n^{.5} a_p^{.5} + b_5 a_n^{.5} a_k^{.5} + b_6 a_p^{.5} a_k^{.5} + b_7 a_n + b_8 a_p + b_9 a_k$

Assumptions

The use of $a_n X = N$ type relationships for the above derivations requires the assumptions:

1. The analysis a_n gives equivalent measures of nutrient content and availability to plants in both the fertilizer X and the fertilizer which was used to estimate the yield function.
2. The effects of the fertilizers on crop production are due solely to their nutrient contents as measured by the analyses and not to other fertilizer constituents or features.

The assumption 1 may not be justified when the nutrients have very different chemical forms for the different fertilizers as for example when the nutrient function is derived using fertilizers with water soluble sources of the nutrients and the alternative fertilizers have nutrients in less readily available forms, for example when the "fertilizer" is an animal manure. The assumption 2 also may not be justified when the alternative fertilizer contains additional nutrients or toxic substances, or has a different physical form to the fertilizers used to obtain the nutrient function.

Substitution rate

In the absence of a satisfactory analytical procedure to obtain fertilizer analyses a_n, etc. for the above procedures, the equivalent in the form of substitution rates can be estimated from experimental data to indicate the nutrient content of a fertilizer relative to some standard fertilizer. For example given a mineral rock phosphate, a substitution rate s_p can be estimated to indicate its equivalent content of phosphorus to that in a standard phosphate fertilizer. Thus, following the same procedure as with an analysis a_p, equivalent rates of application of P in the standard and alternative fertilizer are calculated by $X = P / s_p$ where X is rate of application of the alternative fertilizer as kg fertilizer/ha and P is the rate of application of phosphorus with the standard fertilizer. The value for s_p may be estimated from data obtained from an experiment with treatment applications of a standard P fertilizer at rates represented as kg P/ha and corresponding rates of application of the alternative fertilizer, using rates kg X/ha chosen as best possible to be equivalent, using for example analyses a_p obtained by what seems to be the most appropriate fertilizer analysis procedure. The data from the two fertilizers can be combined to estimate a yield function of form

$$Y = b_0 + b_1 (s_p X)^{.5} + b_2 s_p X$$

where s_p is a substitution rate with the value $s_p = 1$ for the standard fertilizer with rates expressed as kg P/ha and some other value for the alternative fertilizer. Values for the coefficients and the substitution rate s_p for the alternative fertilizer are estimated by the usual regression procedure, using successive approximations of values for s_p for the alternative fertilizer to obtain a best value that minimizes the residual mean square in the regression analysis of variance. If analyses a_p by some procedure give a direct measure of an equivalent nutrient content in the alternative fertilizer, then $s_p = a_p$. Estimates of s_p can thus be used to evaluate methods of fertilizer analysis. For an example use of this procedure with alternative phosphorus fertilizers see Colwell and Goedert (1988).

3.7 Optimal rates for manures

The same procedures to the above can be used to calculate optimal rates of manures using the composition of the manure to calculate "fertilizer" functions. Moreover the procedure can be extended to calculate optimal simultaneous rate of manures and fertilizers or of optimal rate of application of fertilizers in the presence of a specified application rate of the manure. Such calculations are likely to give doubtful estimates however unless the above assumption 1 can be justified, given the very different chemical forms of nutrients in manures to that in single nutrient fertilizers. Yield functions for manures are thus better estimated from experiments with manure rates as treatments so that functions of the form $Y = b_0 + b_1 M^{.5} + b_2 M$ can be estimated directly. Such experiments with manures can be extended by adding treatments for nutrient rates so that for example a manure-fertilizer-yield function

$$Y = b_0 + b_1 M^{.5} + b_2 P^{.5} + b_3 K^{.5} + b_4 (MP)^{.5} + b_5 (MK)^{.5} + b_6 (PK)^{.5}$$
$$+ b_7 M + b_8 P + b_9 K$$

can be estimated from data provided from an experiment with factorial combination of treatment rates for the manure M and the nutrients P and K. Optimal simultaneous rates of application of the manure M and the nutrients P and K can then be calculated from functions of this form using the formula (3.11).

Perhaps more usefully for studies with manures, optimal supplemental rates of application of fertilizer nutrients can be calculated for a nominated rate of application of the manure using the procedure described above for non-ideal functions. For example the above function for M, P and K can be used to derive two nutrient functions with the form

$$Y = c_0 + c_1 P^{.5} + c_2 K^{.5} + c_3 (PK)^{.5} + c_4 P + c_5 K$$

for nominated values for M and this function can then be used to calculate optimal rates of P and K by (3.10) to supplement the nominated applications of manure. Corresponding procedures may be used to derive functions for the calculation of supplemental rates of application of one or three nutrients with the optimal rate formulas (3.9) and (3.11).

3.8 Optimal rates for various types of crop

The procedures for calculating optimal rates considered so far have all been for a simple cropping system in which fertilizer is purchased and applied to an individual crop and in which a single economic return is received after the crop has been grown and sold. Fertilizers are also used in more complex situations so that these basic procedures have to be modified or adapted. The following examples are for some such situations.

Yield and quality

Fertilizer applications to some crops affect their economic value by affecting their quality as well as their yields so that effects on both quality and quantity must be provided for when calculating optimal rates. For example with a vegetable crop, fertilizer applications may affect both the size and the yield of the vegetables and since the economic value of some sizes is greater than of others, rates have to be estimated to optimize effects on the quality (size) *and* the quantity (yield) of crop product. Thus with a beetroot crop that is to be sold for canning, fertilizer affects both the size and the yield of the product and because the canning industry prefers a particular size, the value of that size is much greater than that of any other, larger or smaller.

Optimal rates in this situation are calculated by separating the yield function into components corresponding to each quality class and then calculating the rate that gives the maximum total of the profits from the different components. Fertilizer experiments are carried out as with other crops but then the harvested crop is separated into quality grades so that the yield data for each grade can be used to estimate yield functions for each grade. If for example the yield data from a single nutrient experiment is separated into small, medium and large size grades then yield functions may be estimated for each grade so that instead of simply estimating a function for yield

$$Y = b_0 + b_1 N^{.5} + b_2 N$$

functions are estimated for each quality grade

$$Y_S = c_0 + c_1 N^{.5} + c_2 N$$

$$Y_M = d_0 + d_1 N^{.5} + d_2 N$$

$$Y_L = e_0 \quad e_1 N^{.5} + e_2 N$$

where Y_S, Y_M and Y_L are yields for the small, medium and large size grades. If the economic values for each of these grades are respectively V_S, V_M and V_L, the profit function for the harvest becomes

$$\Pi = V_S Y_S + V_M Y_M + V_L Y_L - C_n (1+R)^t - Q \tag{3.16}$$

and the optimal rate equation corresponding to (3.9) become

$$N = \left(\frac{0 \cdot 5(V_S c_1 + V_M d_1 + V_L e_1)}{C_n (1+R)^t - (V_S c_2 + V_M d_2 + V_L e_2)} \right)^2 \tag{3.17}$$

Here N is optimal because it produces the maximum profit as determined by the total of the values for the yield components Y_S, Y_M and Y_L.

Corresponding profit equations and optimal rate formulas can be derived similarly for other types of quality grades and also for multinutrient functions.

Crops with several harvests

The economic return from a fertilizer application to some crops is obtained from its effects on several successive harvests, as with perennial crops. For example fertilizer may be applied to sugar cane at the time of planting and returns then obtained from that crop and several following ratoon crops. Similarly fertilizer may be applied to a forest at the time of planting and returns received over a period of many years with each of a series of successive selective tree fellings. A similar more difficult example is for a fertilizer application to pasture, the successive returns being obtained by the sale of animal products over a period of perhaps several years. With such crops optimal rates must be calculated from a profit function with time discounting for each of the successive harvests. This is done conveniently by adjusting all returns to their value at the time of purchase of fertilizer as described for (2.14). Thus if the effects of a nutrient application at rate N on n successive harvests are represented by the n yield functions

$Y_1, Y_2, \ldots Y_n$ the profit function with all values adjusted to the time of the commencement of the investment with the purchase of fertilizer is

$$\Pi_0 = \frac{VY_1}{(1+R)^{t_1}} + \frac{VY_2}{(1+R)^{t_2}} + \cdots + \frac{VY_n}{(1+R)^{t_n}} - C_n N - Q_0 \tag{3.18}$$

where $t_1, t_2, \ldots t_n$ are the interest conversions corresponding to the n time periods for the respective harvests.

The calculation formulas (3.9) to (3.11) can be adapted for this type of function. Thus for n single nutrient functions

$$Y_1 = a_0 + a_1 N^{.5} + a_2 N$$

$$Y_2 = b_0 + b_1 N^{.5} + b_2 N$$

$$\ldots$$

$$Y_n = c_0 + c_1 N^{.5} + c_2 N$$

the profit function becomes

$$\Pi_0 = B_0 + B_1 N^{.5} + B_2 N - C_n N - Q$$

where

$$B_0 = V\left(\frac{a_0}{(1+R)^{t_1}} + \frac{b_0}{(1+R)^{t_2}} + \cdots + \frac{c_0}{(1+R)^{t_n}}\right)$$

$$B_1 = V\left(\frac{a_1}{(1+R)^{t_1}} + \frac{b_1}{(1+R)^{t_2}} + \cdots + \frac{c_1}{(1+R)^{t_n}}\right)$$

$$B_2 = V\left(\frac{a_2}{(1+R)^{t_1}} + \frac{b_2}{(1+R)^{t_2}} + \cdots + \frac{c_2}{(1+R)^{t_n}}\right)$$

and the optimal rate equation derived by $d\Pi_0 / dN = 0$ is

$$N = \left(\frac{0.5B_1}{C_n - B_2}\right)^2 \tag{3.20}$$

Profit functions may be developed further with provision for different values for successive harvests as when trees are harvested from a developing forest or when the value of animal production varies with time. Denoting the values of the successive harvests by $V_1, V_2, \ldots V_n$ the above formulas for B_0, B_1 and B_2 become

$$B_0 = \left(\frac{V_1 a_0}{(1+R)^{t_1}} + \frac{V_2 b_0}{(1+R)^{t_2}} + \cdots + \frac{V_n c_0}{(1+R)^{t_n}} \right)$$

$$B_1 = \left(\frac{V_1 a_1}{(1+R)^{t_1}} + \frac{V_2 b_1}{(1+R)^{t_2}} + \cdots + \frac{V_n c_1}{(1+R)^{t_n}} \right)$$

$$B_2 = \left(\frac{V_1 a_2}{(1+R)^{t_1}} + \frac{V_2 b_2}{(1+R)^{t_2}} + \cdots + \frac{V_n c_2}{(1+R)^{t_n}} \right)$$

and the optimal rate equation has, again, the form (3.20).

Crops with successive fertilizer applications

With some crops a nutrient may be applied in successive applications of the fertilizer. For example a nitrogen fertilizer may be applied to a maize crop first at the time of sowing and then again later at the time of tillering. The yield functions and calculation procedures for simultaneous applications of different nutrients can be adapted to calculate optimal rates for the successive applications as with the following examples where application rates are represented by N_s for nutrient applied at the time of sowing and N_t for nutrient applied at the time of crop tillering. The N_s and N_t may be defined of course for other times of application and the procedure extended, at least in theory, for more than two successive applications. For all such calculations the simultaneous rate procedures are particularly appropriate because large negative interaction effects can be expected to be produced by the successive applications.

If the effects on crop production of two successive applications of a nutrient N at the rates N_s and N_t can be represented by the function

$$Y = b_0 + b_1 N_s^{.5} + b_2 N_t^{.5} + b_3 (N_s N_t)^{.5} + b_4 N_s + b_5 N_t \tag{3.21}$$

then optimal rates can be calculated by the formulas (3.10) for simultaneous applications of the two nutrients N and P, substituting N_s and N_t for N and P and assuming the same value for the time variable t, that is that the fertilizer is purchased at the same time for the successive applications.

Similarly if the effects of two successive applications of a nutrient N at the rates N_s and N_t and a single application of a nutrient P at the rate P_s can be represented by the function

$$Y = b_0 + b_1 N_s^{.5} + b_2 N_t^{.5} + b_3 P_s^{.5} + b_4 (N_s N_t)^{.5} + b_5 (N_s P_s)^{.5} + b_6 (N_t P_s)^{.5}$$
$$+ b_7 N_s + b_8 N_t + b_9 P_s \quad (3.22)$$

then optimal rates can be calculated by the formulas (3.11), substituting respectively N_s , N_t and P_s for N, P and K.

All such calculations depend on the functions having the ideal diminishing form with respect to each nutrient variable so that checks for validity as described for the simultaneous rate calculations are important. "Serial" type designs can be used for experiments to estimate functions (3.21) and (3.22) as described in chapter 7.

3.9 Programs of fertilizer application

If soil nutrient deficiencies are severe, programs of fertilizer application can be envisaged for the development of soil fertility and agriculture in which high rates of fertilizer application are used for first crops and progressively lower rates for following crops as soil nutrient levels are increased by the accumulation of residues from the successive applications. Such programs can be expected to lead eventually to a soil nutrient level that only needs to be maintained with maintenance rates of fertilizer application to replace the nutrients removed with cropping and to sustain the agriculture and soil fertility at an optimal level. Ideally the program of fertilizer applications will be chosen to provide maximum economic returns for the development phase so that it may be described as an optimal fertilizing program. Ideally also the fertilizer rates for the subsequent sustenance phase will produce maximum economic returns so that it may be described as an optimal maintenance rate and the soil nutrient level thus established as being at an optimal level of soil fertility.

Optimal programs of fertilizer application can be calculated, in theory at least, if yield functions can be estimated for each of the successive crops of a cropping system and used to calculate the rates which will give the maximum economic return for the development period. Thus if a series of n successive crops have the yield functions Y_1, Y_2, ... Y_n then the total economic return, time discounted to the commencement of the program, for a program of fertilizer applications of the nutrient N at the rates N_1, N_2, ... N_n will be

$$\sum_{i=1}^{n} \Pi_i = \frac{V_1 Y_1}{(1+R)^{t_1}} - C_n N_1 + \frac{V_2 Y_2}{(1+R)^{t_2}} - \frac{C_n N_2}{(1+R)^{s_2}} + \cdots$$

$$+ \frac{V_n Y_n}{(1+R)^{t_n}} - \frac{C_n N_n}{(1+R)^{s_n}} - \sum Q \quad (3.23)$$

where V_i is the value of the i th crop, C_n is the cost of the fertilizer nutrient (assumed constant), t_i is the number of interest conversion periods to the return from crop i, s_i is the interest conversion periods to the investment in the fertilizer at rate N_i (s_1 is zero) and ΣQ is total fixed costs. The program N_1, N_2, ... N_n will be optimal if it produces a greater total profit than any alternative program.

The difficulty of calculating an optimal fertilizing program is of course the practical one of obtaining estimates of the successive yield functions with appropriate adjustments for the increases in soil nutrient levels produced by residues of the preceding fertilizer applications. Moreover the problem will be more difficult where there are several soil nutrient deficiencies and simultaneous applications of several nutrients. The equation (3.21) thus only serves to indicate the nature of optimal programs and the nature of a research challenge. Designs for experiments to estimate such programs are discussed in Chapter 7.

The problem of estimating optimal maintenance rates is very much simpler when fertilizer is applied only to maintain soil nutrients at an optimal level since then there is no need to allow for residue effects on following crops. Such situations may be expected in well established agricultural regions where traditional farm practices have developed a sustainable agricultural system with regular applications of fertilizer. In such situations simple fertilizer experiments are needed only to provide bases for adjusting existing rates towards optimal maintenance rates to suit current economic conditions.

3.10 Margins and ridge lines

Two terms that have been fairly widely used relating to the calculation of optimal fertilizer rates are mentioned in an attempt to avoid any confusion relating to the present descriptions.

Margin The amount of crop produced by farming is described as crop production or yield and the increase in yield or production due to some practice such as the application of fertilizer is termed, logically, the increase or response of production or yield. These simple meanings can be confused by the unnecessary use of the word *margin*. This word has important and specific uses in economics, as for example to refer to the minimum return below which an enterprise becomes unprofitable, but it has also been used to simply mean yield, production or response.

The word is also commonly used to refer to an edge or limit so that given a functional relationship with a diminishing rate of return, the variable R in the profit equation $\Pi = VY - C_n N(1+R)^t - Q$ may be described as *marginal rate of return*. It seems simpler however to describe R as an interest rate or as the rate of return for alternative investments.

Ridge lines These have been defined and used for the calculation of optimal rates of application of nutrients with multinutrient yield functions apparently to avoid the need to use partial derivatives. Thus given a multinutrient function $\Pi = f(N, P, \cdots, S)$, a ridge line for the nutrient rate N is calculated by determining series of values for Π and N such that $d\Pi/dN = 0$ for sets of values for the other variables, $P, \ldots, S$. By repeating this procedure to obtain ridge lines for each of the variables, the optimal simultaneous rate is defined by the values of $N, P, \ldots, S$ at the intersection of all of the ridge lines. The calculation of simultaneous optimal rates by the standard differential calculus procedure with the solution of equations obtained by $\partial\Pi/\partial N = \partial\Pi/\partial P = \ldots = \partial\Pi/\partial S = 0$ as with the equations (3.10) and (3.11) is much more direct however, and much simpler in that it does not require the tedious calculation of these so called ridge lines.

Chapter 4

Accuracy of Estimates from Fertilizer-Yield Functions

4.1 Introduction

Mathematical functions relating crop yields to fertilizer application rates are estimated from experimental data by fitting regressions and the estimated functions may then be used to calculate crop yields, profits and optimal fertilizer application rates with mathematical precision. Obviously however, the accuracy of such calculated values is limited by the accuracy of the estimated regressions and, less obviously, by the accuracy with which the regressions represent actual relationships between fertilizer rate and yield. The degree of accuracy or precision to be associated with calculated values needs to be indicated then, particularly to dispel any unwarranted aura of confidence, or scepticism, that can be so easily produced by esoteric computations and impressive computer "print-outs". Unfortunately there is no simple single measure to indicate the accuracy of the values that can be calculated from regressions. Rather, various statistical procedures are used to calculate values that indicate particular aspects of accuracy. It is important then to understand the aspect of accuracy represented by these various values, and their nature, purpose and limitations, to avoid producing false impressions of accuracy. It is not sufficient for example to gauge the accuracy of calculated values from the magnitude of R^2 values or from the statistical significance for some quantity, as commonly indicated with the star ratings *, ** or ***. In particular it is important to understand the nature of the following indices of accuracy that are commonly calculated for regressions and regression analyses of variance, and for present purposes, what they indicate about the accuracy of the various values that are calculated from regression estimates of functions for the fertilizer-yield relationship.

Error variance

Estimates of error variance by the residual mean square of regression analyses of variance indicate the amount of the variation that is not represented or explained by regressions and as such serve as valuable indices of their accuracy. However they do not necessarily give reliable indications of the accuracy of regression estimates that are calculated for values of the regressor variables between those of the data, by interpolation, or beyond the range of the data or data combinations, by extrapolation. More importantly, with regressions for fertilizer-yield functions, they do not indicate the accuracy of estimates for conditions other than those that prevailed for the experiments that produced the data for the regressions and consequently of estimates for crops in regions supposedly represented by the experiments.

Tests of significance for treatment effects

Standard tests of significance for treatment effects in analyses of variance are based on comparisons of the magnitude of treatment effects with error effects as estimated by the error variance. For example when treatment effects are not much different from effects attributable to error they are rated as non significant. The tests do not give reliable and direct measures of accuracy however because they are based on such comparisons rather than direct estimates of error. Thus small non significant effects may be estimated just as accurately as large highly significant effects.

Coefficient of variation

Coefficients of variation provide convenient, scaleless measures of the error associated with regressions but can mislead because they vary inversely with the magnitude of the data mean.

R^2

R^2 values provide a convenient scaleless measure of the results of regression analyses of variance but do not give reliable measures of accuracy because they represent a ratio, that is the ratio of the magnitude of the data variability represented by regressions to the overall variability of the data from which they have been estimated.

R_a^2

R_a^2 values, or R^2 *values adjusted for degrees of freedom*, are estimates of the proportion of the variance of the dependent variable of a regression accounted for by the regression and as such they are often more informative than R^2 values. They also have a limit on their usefulness corresponding to that for R^2 values however because they also are values for a ratio.

Confidence intervals

Standard confidence intervals for regressions indicate the accuracy of estimates of the dependent variable by the regressions but may not indicate the accuracy of other values that may also be calculated from the regressions. Thus with regressions for fertilizer yield functions they may be used to indicate the accuracy of estimates of yield calculated from the regressions but not the accuracy of estimates of optimal fertilizer rates calculated from the regressions. Other types of confidence interval can be calculated however for other values, in particular for optimal rates.

4.2 Accuracy of estimates by regressions

Example regression

The use of the above common statistical quantities as indices of accuracy is illustrated with the fertilizer-yield function

$$Y = 2809.1 + 423.80P^{\cdot 5} - 21.783P - 17.0L_b \qquad (4.1)$$

estimated by the usual regression procedure from the experiment data in Table 4.1. The data are from a simple fertilizer experiment with wheat in which there were five treatment rates for a P fertilizer, all replicated in two blocks so that the regression gives a convenient representation of the results of the experiment in the form of a mathematical equation, with Y for yield (kg wheat/ha), P for nutrient application rate (kg P/ha) and L_b for a dummy variable with values -1 and $+1$ for location in blocks 1 and 2 respectively. Estimates of yield for the experiment site considered as a whole can be calculated from this regression by substituting the mean value $L_b = 0$, that is by simply dropping the term $-17.0L_b$ from the regression. Profit functions can be derived from the regression and profits and optimal fertilizer rates

Table 4.1 Yield data from a fertilizer experiment in two blocks.

Block	Treatments, kg P/ha				
	0	5	10	20	40
1	2590	3810	3920	4270	4770
2	2990	3840	3840	4210	4500
Mean	2790	3825	3880	4240	4635

calculated using nominated values for economic variables as described in the preceding chapters. For the example the values $V = \$0.08$ per kg wheat, $C_p = \$1.92$ per kg P, $R = 0.02$ for interest rate and $t = 1$ for interest conversion period have been chosen giving the economic variable value $E_p = \dfrac{(1.92)(1+0.2)^1}{0.08} = 28.8$ and hence the optimal rate

$$P = \left(\frac{0.5b_1}{E_p - b_2}\right)^2 = \left(\frac{(0.5)(423.80)}{28.8 - (-21.783)}\right)^2 = 17.55 \text{ kg P/ha}$$

by the optimal rate equation (3.9).

The procedures illustrated with this example regression for a single nutrient function can be applied similarly with regressions for multinutrient relationships, although in some cases the results are difficult to interpret.

Error variance

If a suitable model has been chosen for a regression, the deviations of data from the estimated regression can be attributed to error effects and since the deviations produce the residual mean square in a regression analysis of variance this value can be used as an estimate of the error variance associated with the data. Since however an estimate obtained this way is derived indirectly, rather than from a direct estimate of "pure" error variance as provided by variations amongst data replicates, it is better called the *residual mean square* rather than the error variance, to cover the possibility that some of the residual sum of squares in the analysis of variance is produced by inadequacies in the

Table 4.2 Analysis of variance for regression (4.1).

Source of variation	Degrees of freedom	Sum of squares	% Sum of squares	Mean square	F ratio
Regression	3	3,795,340	96.12	1,265,113.3	49.58***
Residue	6	153,108	3.88	25,518.0	
Total	9	3,948,450	100.00		

(***, probability<0.001).

model chosen for the regression. If the use of an inappropriate model for a regression contributes to the data deviations, then the residual mean square will over-estimate the true error variance in the data. In such a case the residual mean square in a regression analysis of variance still serves as a measure of accuracy even though it does not estimate the true error variance. For convenience residual mean square is often denoted by RMS.

The residual sum of squares in the regression analysis of variance in Table 4.2 for the example regression (4.1) is 153,108 giving the residual mean square 153,108 ÷ 6 = 25,518 and the F ratio is calculated by the ratio of the regression mean square with this value, F = 1,265,113.3 / 25,518.0 = 49.58. A significance rating for this F value is obtained by treating the residual mean square as an estimate of error variance giving the rating *** for probability < 0.001 since the statistical table value of F for 3, 9 degrees of freedom and probability = 0.001 is F = 13.90. If a corresponding analysis was made with a better model for the relationship, the residual mean square would be smaller, the F ratio greater and the test of significance better. It is common however to use regression residual mean squares as estimates of error variance to obtain significance ratings with the tacit assumption that the regression gives a sufficiently good representation of the true relationship.

If the value for the residual mean square from an analysis of variance of an experiment, RMS = 25,518.0 for the example, can be compared with corresponding values from other similar experiments, then its magnitude can be used as a measure of the relative quality of the experiment. For example if most similar experiments had RMS values greater than 25,518 then the "experimental error" for the experiment would be rated as being low and the experiment as being better than average. Such direct comparisons are however not always possible, as when there have been few similar experiments or when comparisons are wanted for experiments with different types of crop and different units of measurement. For example comparisons of RMS values will not answer questions such as: "Is the quality of fertilizer experiments with wheat in Australia comparable to those with coffee in Brazil?" In any case, the numerical values for RMS are not easy to remember for general comparisons. Consequently alternative scaleless measurements that can be expressed on a percentage scale are much more popular although they also have their limitations.

Coefficient of variation

The problem of comparing estimates of error variance due to the effects of the scale of measurement on the values obtained in analyses of variance has led to the popularity of a scaleless measure, called the coefficient of variation, obtained by dividing the square root of the residual mean square by the data mean,

$$CV = \frac{(RMS)^{.5}}{\overline{Y}} \tag{4.2}$$

where CV is coefficient of variation, RMS is residual mean square or "error variance" and $\overline{Y}$ is the data mean. For the example experiment

$$CV = \frac{(25518)^{.5}}{3855} = 0.041 \quad \text{or} \quad 4.1\%.$$

Such values are scaleless and easily remembered for comparisons with other regressions, perhaps representing treatment effects on different crops with different units of measurement. Thus on the basis of experience with many fertilizer experiments, the ratings in Table 4.3 might be used to evaluate the quality of regressions and hence of experiment data. On this subjective basis, the value CV = 4.1% for the example indicates a very low level of experimental error and a high quality experiment.

There is however an important restriction on the use of the coefficient of variation as a measure of quality. Because CV values are obtained by dividing $(RMS)^{.5}$ by $\overline{Y}$, CV values vary inversely with the magnitude of the data mean and can consequently give misleading impressions of relative levels of error when $\overline{Y}$ values vary widely. Thus experiments for a particular crop with high overall levels of yield will tend to have lower CV values than those with low levels of yield. When

Table 4.3 An evaluation of experimental error for fertilizer experiments on the basis of coefficient of variation (CV).

CV %	Experimental error
>20	Very high
15 - 20	High
10 - 15	Average
5 - 10	Low
<5	Very low

however error variance is proportional to the level of yield, CV values provide a convenient indication of quality.

Tests of significance

Standard analyses of variance are used to obtain tests of significance for treatment effects in experiments and for reasons that are not always warranted, experiments that produce highly significant effects are often regarded as being better than those that produce low or non-significant effects. Although highly significant effects are more likely to be obtained with high quality experiments in which the error variance is low, this does not mean that high significance necessarily indicates low error or conversely that a low significance necessarily indicates a high error. The reason for this is that tests of significance are based on the magnitudes of treatment effects relative to error effects. Thus the popular F test in analyses of variance is the ratio of the variance attributable to the treatments to the variance attributable to error, as represented by the residual mean square, and because it is a ratio, significance ratings are determined by the magnitude of treatment effects as well as by the magnitude of error effects. For example if fertilizer experiments are carried out on a soil with severe nutrient deficiencies, even poor quality experiments with large error effects may produce highly significant results, simply because the treatment effects are large and obvious. In contrast, if experiments are carried out on a soil with no nutrient deficiencies, even high quality experiments with very small error effects will produce non-significant results, because there are no treatment effects. The accurate determination of small treatment effects which may also be non-significant, or of a nil effect which will certainly be non-significant, can be as important as large effects from a research point of view. Tests of significance should consequently not be used to indicate quality.

R^2 values

Similarly the popular R^2 value with the impressive name *coefficient of determination,* obtained with regression analyses of variance, is an unreliable indicator of quality because it also is a ratio, this time of the regression sum of squares to the total sum of squares. Thus for the example, R^2 = 3,795,340 / 3,948,450 = 96.12% (Table 4.2), this ratio simply indicates the magnitude of effects that are accounted for by regression relative to the overall variability of the data and as such may or may not indicate the quality of the data. Low R^2 values may be obtained both in situations where there are large error effects and where there are small error effects but also little or no treatment effects to be explained by the regression.

An important example of the inappropriate use of R^2 values as an indication of the quality is seen with values obtained for regressions with different ranges of treatment values. For example the inclusion of the zero rate as a treatment in fertilizer experiments helps to produce high R^2 values because it produces a wider range of yield data but this does not necessarily mean that it will produce better estimates of the fertilizer-yield relationship for the important range of fertilizer rates in the vicinity of the optimal rate. For example if optimal fertilizer rates occur in the range 5 to 40 kg N/ha, then although experiments with the treatment rates 0, 20, 40 and 80 will tend to produce higher R^2 values than experiments with treatments 5, 10, 20, 40, the functions estimated with the more restricted treatment range are likely to give more accurate estimates of the actual fertilizer-yield relationship in the important range 5 to 40 and hence better estimates of optimal rates, because the model used for the regression is better able to represent the actual relationship with a restricted range of data. This would be seen if the residual mean squares were compared but the comparison of R^2 values would be likely to give the opposite impression.

R_a^2 values

The development of multiple regressions often requires the comparison of alternative regressions, with different numbers of regressor variables and in such cases R^2 values are misleading because their values always increase with the addition of a regressor variable, whether or not the addition produces a better representation of the relationship as indicated by the effect of the addition on the magnitude of the residual mean square in analyses of variance. R_a^2 values, or as they are sometimes called, "R^2 values adjusted for degrees of freedom", are more useful for such comparisons because they represent the proportion of variance explained by regressions, irrespective of the number of regressor variables, and their magnitudes decrease if the addition of a variable increases the residual mean square. Values are calculated by

$$R_a^2 = 1 - \frac{RMS}{VAR} \tag{4.3}$$

where *RMS* is the residual mean square in the regression analysis of variance and *VAR* is the variance of the dependent variable, given by $TSS/(n\text{-}1)$ where *TSS* is total sum of squares of deviates and n is the number of observations.

R_a^2 values relate to R^2 values by

$$R_a^2 = 1 - \frac{(1 - R^2)(n - 1)}{(n - k - 1)}$$

where n = number of observations and k = number of regressor variables. R_a^2 values are always less than those for corresponding R^2 values and moreover will decrease with the addition of a variable to a regression when the increase in the regression sum of squares is less than the residual mean square whereas R^2 values always increase with the addition of a variable. In extreme cases when the residual mean square is greater than the variance of the dependent variable Y, R_a^2 values are negative.

Confidence intervals

A popular and convenient procedure for showing the accuracy of estimates obtained from a regression equation is to calculate confidence intervals as with the graph of the example regression in Fig. 4.1. This interval is however only for estimates of the dependent variable, Y, and other intervals must be calculated for other values that may be estimated from regressions. In particular intervals for regression estimates of yield as in Fig. 4.1 do not indicate the accuracy of estimates of optimal fertilizer rates.

Intervals for estimates of yield

The most common confidence interval given with regressions is for estimates of the dependent variable as illustrated in Fig. 4.1. If $\hat{Y}$ (termed "Y hat") is used to denote an estimate of Y from the regression,

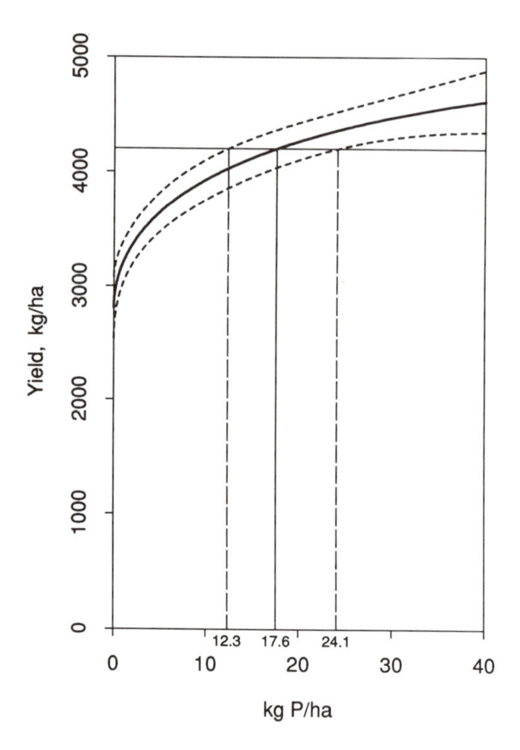

Fig. 4.1 Yield function (4.1) and 95% confidence limits for yield. The optimal rate is P = 17.55.

the interval is defined by $\hat{Y} \pm ut$ where u is the standard error of $\hat{Y}$ and t is a value for the "t-distribution" for a chosen level of probability and the degrees of freedom for the error variance, as listed in statistical tables. Given a regression

$$Y = b_0 + b_1X_1 + b_2X_2 + \cdots + b_kX_k$$

the variance of an estimate $\hat{Y}$ for a particular set of values of the k regressor variables X_1, X_2, ... X_k is calculated by the matrix multiplication

$$\mathrm{Var}(\hat{Y}) = s^2\begin{bmatrix}1 & X_1 & X_2 & \cdots & X_k\end{bmatrix}\begin{bmatrix}c_{00} & c_{01} & c_{02} & \cdots & c_{0k}\\ c_{10} & c_{11} & c_{12} & \cdots & c_{1k}\\ c_{20} & c_{21} & c_{22} & \cdots & c_{2k}\\ \vdots & \vdots & \vdots & \ddots & \cdots\\ c_{k0} & c_{k1} & c_{k2} & \cdots & c_{kk}\end{bmatrix}\begin{bmatrix}1\\ X_1\\ X_2\\ \vdots\\ X_k\end{bmatrix} \qquad (4.4)$$

where s^2 is the estimate of error variance provided by the residual mean square of the regression analysis of variance and the c_{ij} are elements of the inverse of the variance-covariance matrix $[\mathbf{XX}']^{-1}$ that was used to calculate the estimates of the regression coefficients. The standard error for $\hat{Y}$ is the square root of this variance, $u = \sqrt{\mathrm{Var}(\hat{Y})}$. Thus for the example, the estimate of the yield $\hat{Y} = 4202$ obtained for the estimated optimal rate $P = 17.55$ is calculated by substituting this rate in the regression and the 95% confidence interval for this estimate $\hat{Y} = 4202 \pm 165$ is calculated from the inverse matrix for the regression, $s^2 = 25518$ from the analysis of variance in Table 4.2 and $t = 2.4469$ for 95% probability and 6 degrees of freedom as given in statistical tables. The graphs for the yield function and the 95% confidence intervals in Fig. 4.1 have been obtained by making such calculations with the range $P = 0$ to 40 and $B_L = 0$, corresponding to the range of the P treatment rates and the mean for the block dummy variable.

In simple terms a 95% confidence interval for a regression obtained in this way from a set of experimental data is often described, simply but not strictly accurately, as the interval in which 95% of the regressions would occur that would be obtained by many repetitions of an experiment under identical experimental conditions. More accurately but less comprehensibly, if 95% confidence intervals were obtained for a large number of regressions estimated from data obtained by many repetitions of an experiment, then 95% of these intervals would contain the mean or "true" regression. In practice such descriptions do not seem to help much in evaluating the accuracy of regression estimates and in general it is sufficient to understand that the interval is simply an indication of accuracy, small intervals for $\hat{Y}$ being better than large.

Unfortunately such explanations are often forgotten with a consequent misuse of confidence intervals so that it is worth emphasizing some of the things that they do not show:

1. Confidence intervals for a regression estimate of a fertilizer-yield relationship, obtained from the data of a particular fertilizer experiment, do not indicate the accuracy or reliability of values calculated from the regression as a guide for fertilizer use in the region represented by the experiment. In typical variable regions many factors affect the relationship and the experimental estimate and confidence interval apply only to the particular set of conditions that happened to prevail for the experiment.
2. The regression estimate of yield is at the centre of the confidence interval and it is the *best* estimate from the experimental data for

the assumed model form of the relationship. Other estimates become progressively less likely with increase in deviation from this estimate. Consequently confidence intervals do not indicate ranges of equally likely estimates and in particular it is not correct to regard estimates near the lower boundary as being preferable because they are more conservative.

3. Confidence intervals for estimates of yield do not indicate the accuracy of estimates of optimal fertilizer rates. Thus the interval for the estimate of the yield $\hat{Y}$ = 4202 ± 165 at the optimal rate $P' =$ 17.55 in the example does not indicate the accuracy of this calculated value as an estimate of the optimal rate. Confidence intervals for optimal rates are obtained by a basically different procedure as described below.

Intervals for yield estimates from multinutrient functions

The equation (4.4) can also be used to calculate confidence intervals for multinutrient functions but the intervals are difficult to illustrate with graphs corresponding to those in Fig. 4.1. This is because graphical representations of yield functions for n nutrients require $n + 1$ dimensional diagrams and confidence intervals require correspondingly $n + 1$ dimensional diagrams. Such diagrams are difficult to construct and to comprehend even in three dimensions for 2-nutrient functions. Consequently it seems pointless to attempt to indicate accuracy of estimates from multinutrient regressions with confidence intervals and better to simply note that standard errors of estimates $\hat{Y}$ increase with the error variance of the regressions.

Intervals for estimates of profit

The profit function

$$\Pi = VY - C_n N(1+R)^t - Q \qquad (4.5)$$

is derived from a single nutrient yield function $Y = f(N)$ where N is nutrient application rate and V, C_n , t, R and Q are economic constants and estimates of the profit, $\hat{\Pi}$, are calculated by first calculating values of $\hat{Y}$ from the yield function and then substituting in this equation. Confidence intervals for estimates of profit can be obtained in the same way by substituting interval values for $\hat{Y}$ in the profit equation. This is illustrated in Fig. 4.2 with the profit function and confidence intervals derived from the yield function and intervals of Fig. 4.1 with the

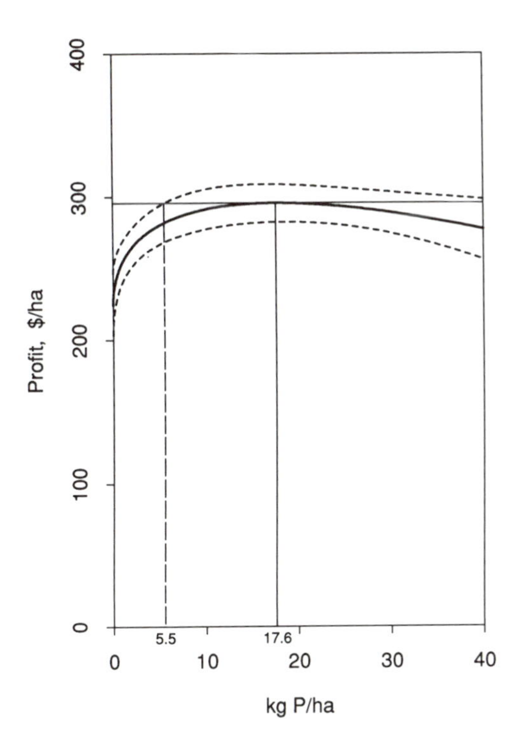

Fig. 4.2 Profit function and 95% confidence intervals derived from the example regression of Fig. 4.1. In this case it is not possible to derive inverse confidence intervals for the rate P = 17.55 kg P/ha producing the maximum profit.

economic values used previously, V = \$0.08, C_p = \$1.92, R = 0.2, t = 1 and omitting a value for fixed costs Q. In particular the function has the estimated maximum and confidence interval $\hat{\Pi}$ = \$295.74 ± 13.21 for the optimal rate P = 17.55 kg P/ha. Again, as with the interval for $\hat{Y}$, this confidence interval for the estimated profit at the optimal rate does not indicate the accuracy of the estimate of the optimal rate.

Inverse confidence intervals

Regression equations can be used "inversely" to calculate values for the regressor variables that produce nominated values for the dependent variable and the relative accuracy of such calculated values can *sometimes* be shown by deriving inverse confidence intervals from the confidence intervals for the dependent variable. The procedure is illustrated in Fig. 4.1 with the horizontal line for Y = 4202 intersecting the graphs for the 95% confidence intervals at P = 12.3 and 24.1. The inverse estimate $\hat{P}$ = 17.55 can be calculated for Y = 4202 from the regression and these intersect values give the confidence interval $12.3 < \hat{P} < 24.1$. The rationale for this procedure is that all 95% confidence intervals for estimates of Y for values of P in the range 12.3 to 24.1 contain the value Y = 4202. Note incidentally that the inverse interval is asymmetric and consequently must be indicated with < symbols rather than ±.

This procedure for obtaining inverse confidence intervals can only be used with certain forms of regression relationship and problems are often encountered with regressions. Draper and Smith (1981) comment for example that "inverse estimation is, typically, not of much practical value unless the regression is well determined". Such a problem with the procedure is illustrated in Fig. 4.2 with an attempt to obtain inverse limits for the estimate of the application rate $\hat{P}$ = 17.55 that produces the maximum profit Π = 295.74. In this case the horizontal line for Π = 295.74 can only intersect the graph for the upper limit. The problem in this case is not very important since it is obvious from an inspection of the graphs that a wide range of application rates are likely to produce profits close to the maximum of Π = 295.74 or in other words, that the profit function has a high economic elasticity (section 3.3) in the vicinity of its maximum.

Intervals for estimates of the optimal rate

The intervals described so far are for estimates of yield and profit calculated from regression equations for Y and Π using the variances and covariances associated with the estimates of the equation coefficients and it is important to understand that because estimates of $\hat{Y}$ and $\hat{\Pi}$ are obviously not estimates of the optimal rate, the intervals cannot indicate the accuracy of estimates of the optimal rate. Intervals for the optimal rate require corresponding equations and information about the variances and covariances of the coefficients in the optimal rate equations. In particular, if the optimal rate equations can be arranged in a linear form with respect to the coefficients, the same type of procedure can be used as described above with (4.4) for Var($\hat{Y}$). Optimal rate equations for square root quadratic functions are convenient in this respect. Thus given a regression with the linear form corresponding to that of the example

$$Y = b_0 + b_1 P^{.5} + b_2 P + b_3 B_L$$

the optimal rate is calculated by solution of the equation

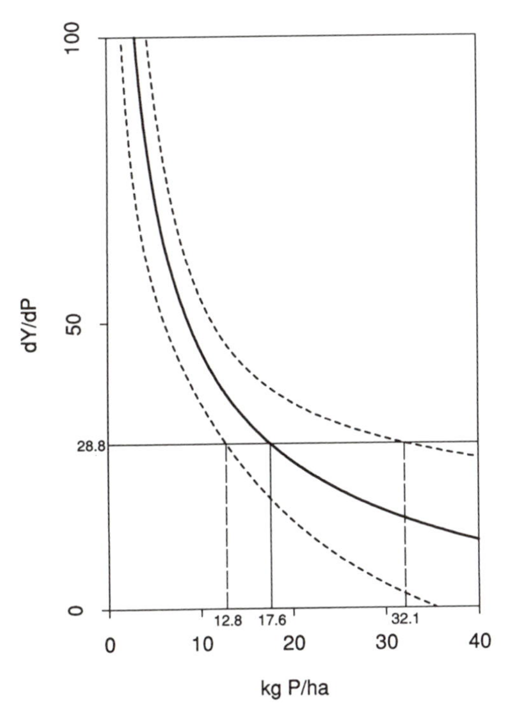

Fig. 4.3 Graphs for the slope d*Y*/d*P* for the example function of Fig. 4.1 and 95% confidence intervals. The inverse interval for P = 17.55 gives a confidence interval for the optimal rate.

$$\frac{dY}{dP} = \frac{b_1}{2P^{.5}} + b_2 = E_{\text{p}} \tag{4.6}$$

where $E_p = \dfrac{C_p(1+R)^t}{V}$. Since $\dfrac{b_1}{2P^{.5}} + b_2$ has a linear form with respect to the coefficients b_1 and b_2, the variance of dY/dP can be calculated directly by

$$\text{Var}\left(\frac{dY}{dP}\right) = \frac{c_{11}}{4P} + \frac{c_{12}}{P^{.5}} + c_{22} \tag{4.7}$$

where c_{11}, c_{12} and c_{22} are elements from the variance-covariance matrix for the coefficients b_1 and b_2. The variance calculated in this way can be used to calculate confidence intervals for dY/dP as illustrated for the example regression in Fig. 4.3 and inverse confidence intervals can then be calculated for values of P that give nominated values for the slope dY/dP. If the slope $dY/dP = E_p$ is nominated, the interval obtained in this way is for *the optimal rate.*

The calculation of confidence intervals for estimates of optimal rates becomes more difficult when they involve non-linear expressions that do not allow straightforward calculations of variances from variance-covariance matrices as with linear expressions. The problem does not arise with the $Y = b_0 + b_1N^{\cdot 5} + b_2N$ regression model because it is possible to calculate intervals for the linear expression $\frac{b_1}{2P^{\cdot 5}} + b_2$ derived from the non-linear expression $\left(\frac{0 \cdot 5b_1}{E_p - b_2}\right)^2$ derived in turn from the optimal rate equation $dY/dP = E_p$.

For the example, the inverse confidence interval for the optimal value $P = 17.55$ is $12.8 < P < 32.1$. Note that the confidence interval is asymmetric and in particular that it is very different from intervals that might be suggested, erroneously, by intervals for yield or profit with $P = 17.55$ in Figs 4.1 and 4.2.

Corresponding calculations for multinutrient functions are much more difficult and given that the interval regions are also much more difficult to visualize they hardly seem worth the trouble. Thus although in theory confidence regions can be calculated, the relative accuracy of estimates is better assessed from a practical viewpoint in more general terms on the basis of the magnitude of the error variance or coefficient of variation estimated by standard analyses of variance.

Usefulness of confidence intervals

All confidence intervals calculated for regression estimates apply only to the particular regression and to the conditions that produced the data from which the regression was estimated. Thus confidence intervals for estimates of the optimal rate calculated from the data of a fertilizer experiment apply only to the conditions that prevailed for that experiment and do not indicate the accuracy of the estimate for other conditions, such as might exist for other sites, and other years, in the region represented by the experiment. The usefulness of confidence intervals as an indication of the accuracy or reliability of estimates of fertilizer requirements in a region based on the results of individual fertilizer experiments is thus very limited. In simple terms good experiments are better than poor and the more good experiments there are to represent a region, the better the estimates of fertilizer requirements for that region. Confidence intervals provide only one rather limited indication of the usefulness of experiments in this respect.

4.3 Sources of error

The above statistical quantities indicate aspects of accuracy relating to, and *only* relating to, estimates from regression equations. As such they have a very limited value as indicators of the accuracy of estimates of yields, profits, fertilizer requirements and other quantities that may be made for regions from the results of fertilizer experiments because there are other major sources of error not represented by these quantities. It is important then to recognize the sources of error that can contribute to the inaccuracy of estimates based on regressions that have been estimated from experimental data.

Within site error

Ideally experiments are carried out under perfectly uniform conditions so that the data they produce only differ from each other because of the effects of the treatments. With field experiments however, despite all efforts to attain this ideal by the careful choice of uniform sites for the experiments and all attempts to maintain uniform conditions during the progress of the experiments, factors additional to the treatments usually affect the data, producing unexplained and unexpected effects in experimental data, the so-called error effects. The effects may be produced by variations in soil, microclimate, weed infestation, insect damage, disease infection, etc. within the experiment site or inadequacies in experimental technique such as the non-uniform application of treatments and inaccurate measurement of plot yields. Allowance and adjustments for some of these factors can be made, as by partitioning sites into blocks and allocating plots so that block effects can be separated in analyses of variance, but error effects will inevitably remain. Because the error is due to variations of uncontrolled factors within the site it is called *within site error* and the residual mean square in analyses of variance of experimental data are described as an estimate of the *within site error variance*. The quality of experiments, the significance of treatment effects and the accuracy of values that are calculated from regressions are all assessed on the basis of this estimate of within site error variance as described in the preceding sections. Such assessments apply only to estimates for the conditions under which the data were produced, that is for the conditions for the experiments.

Between site error

Another important type of error, termed *between site error,* is associated with the variations between the results from experiments that have been carried out at different sites or times. Although it cannot be measured from the data for individual experiments and is quite different from the within site error considered in this chapter, it is easily confused with the within site error. *Between* site error effects are usually very much greater than *within* site error effects so that failure to distinguish between them can produce very misleading impressions about the accuracy of estimates based on the results of individual experiments when these are used for the region in which they were located.

When experiments are carried out at different sites and in different years, different results are to be expected because of variations in the growing conditions for the different experiments and the variations between experiments that cannot be accounted for by variations of site variables produce the between site error. Thus although experiments should be conducted with all care to minimize within site error, it must be remembered that the growing conditions for a well conducted experiment, as on a research station, may be rather idyllic compared with those that exist under the conditions of practical agriculture in the surrounding region and in any case will differ from those for other locations and growing seasons. Procedures for making estimates for sites in a variable region from the results of experiments in the region and procedures for assessing their accuracy from estimates of between site error variances are described in later chapters.

Errors due to model

Standard procedures for assessing the accuracy of estimates obtained from regressions are made with the tacit assumption that the regressions constitute a direct and unbiased estimate of a functional relationship. If however the mathematical model form chosen for the regression is not suitable for representing the relationship, estimated regressions will produce misleading representations of the relationship with consequent errors, especially for estimates obtained by extrapolation beyond or interpolation between the treatment rates for the experiments from which the regressions have been estimated. Since there is no theory to define the true form of the fertilizer-yield functional relationship, an empirical model must be chosen, that is a model must be chosen on the basis of observation and experience. If a poor choice is made, the model may produce considerable errors in estimates obtained from the regressions and unfortunately there is often little to indicate this. In particular, when there are only a few treatment levels, as is common for practical reasons with fertilizer experiments, the choice of an inappropriate model may have little effect on estimates of error variance even though the model produces large error effects in estimates calculated from regressions, because the estimation procedure is designed to produce as close a fit as possible of the regressions to the experimental data. Thus error effects due to model may not be indicated by the error variance nor consequently by any of the procedures described in this chapter. It is important therefore to know the nature of the error effects that choice of model may produce as described in the following chapter.

Non-random error and data rejection

Standard statistical estimation and analysis of variance procedures are based on a statistical model in which the error effects amongst data are produced by a random error variable which has both positive and negative values and a mean value of zero. When however a few of the data have been affected by some uncontrolled factor, as for example when animals have damaged a few of the plots of an experiment, the error effects are no longer random and estimates of treatment effects and error variance will be misleading unless the affected data are deleted. Estimates obtained from a reduced set of data are of course not as good as those from a complete set and when more than about 5% of the data have to be deleted, it may be preferable to simply reject the whole experiment as a failure.

Although data that have been affected by some mistake or damage should be deleted, it is important that they should not be deleted simply because they look wrong. Such deletions would be equivalent to selecting data to match preconceived ideas about treatment effects and may consequently also produce misleading estimates of actual treatment effects. For example an inspection of the data in Table 4.1 might suggest that the yield value of 2590 for the zero treatment in block 1 is too small because of some unobserved mistake or some unobserved damage that has occurred during the experiment but rejection solely on this basis would produce a bias such that the magnitude of the treatment effects would be underestimated. A test in this regard is that data should only be rejected if it is possible to clearly identify a cause that has produced non-random effects. Thus for example, data should be deleted when there is clear evidence that they have been affected by factors such as animal grazing, roots from a nearby tree or soil compaction from a previously unrecognized track across the experiment site.

Chapter 5

Modelling the Fertilizer-Yield Relationship

5.1 Effects of model on fertilizer requirements

The relationship between yield and fertilizer application rate is represented by fertilizer-yield functions with a general mathematical form, called a model, that is chosen so that functions estimated from experimental data will give realistic mathematical representations of the biological relationships indicated by the data. Thus the function $Y = 2809.1 + 423.80P^{.5} - 21.783P - 17.0L_b$ of the preceding chapter was estimated from the data in Table 4.1 using the model of a square root quadratic

$$Y = b_0 + b_1N^{.5} + b_2N + b_3L_b$$

in which the first part of the model, $b_0 + b_1N^{.5} + b_2N$, is for the effect of nutrient application rate N on yield and the second part, b_3L_b, is for the effect of location in blocks 1 and 2. Similar functions with this same model form may be estimated from other similar sets of data and functions for other types of relationship may be estimated similarly after choosing other suitable models. Thus for example the mathematical form $b_0 + b_1N^{.5} + b_2P^{.5} + b_3(NP)^{.5} + b_4N + b_5P$ might be chosen to represent the effects of rates of application of the two nutrients N and P and the form $b_6L_b + b_7Q_b$ to represent the effects of location in the three blocks of an experiment where L_b and Q_b dummy variables, as defined in chapter 2 (Table 2.2), lead to the model

$$Y = b_0 + b_1N^{.5} + b_2P^{.5} + b_3(NP)^{.5} + b_4N + b_5P + b_6L_b + b_7Q_b$$

for a series of functions to be estimated from the data of a series of experiments with N and P treatment rates and three blocks. There is however a problem. There is no theory or basic knowledge about the way that fertilizer nutrient application rates affect crop yield that can provide an *a priori* definition of the mathematical form for models. Rather models must be chosen empirically, meaning that appropriate models must be assumed on the basis of observation and experience. The problem is important because different models are chosen by different people, all claiming this empirical basis of observation and experience, and the functions estimated with these different models can give very different estimates of yields and profits for nominated application rates, and most importantly, very different estimates of optimal rates. Moreover the effects of model are produced irrespective of the quality of the data and are usually not obvious from standard statistical analyses of variance or tests for goodness-of-fit.

The problem with the choice of model for fertilizer-yield functions is illustrated with the graphs in Fig. 5.1 for the functions

$$Y = 600 + 500P^{.5} - 30P \qquad \text{Optimal } P = 18.1 \qquad (5.1)$$

$$Y = 2675 - 2075\exp(-0.7414P) \qquad \text{Optimal } P = 22.6 \qquad (5.2)$$

$$Y = 2670 + 35(P - 29.57 - |P - 29.57|) \qquad \text{Optimal } P = 29.6 \qquad (5.3)$$

$$Y = 600 + 89.94P - 0.7937P^2 \qquad \text{Optimal } P = 38.5 \qquad (5.4)$$

and in particular by the optimal rates that have been calculated from these functions. Each of these very different regression representations of the relationship and each of the consequent very different estimates of the optimal rate, ranging from 18.1 to 38.5 kg P/ha, has been obtained from the identical set of experimental data in Table 5.1 and the same economic variable value, E_p = 28.8, the differences being produced solely by the choice of model from those listed in Table 5.2. The functions all give identical estimates of yield for the three data application rates 0, 25, 81 kg P/ha and all give identical R^2 values and F ratios for statistical tests of significance because the three parameters for each model are estimated from data with replicates for only three treatment rates. Corresponding contrasts are obtained when functions are estimated from any set of experimental data because of the different mathematical forms that are imposed by the different models. Different R^2 values and F ratios are obtained with different models when there are more data treatment levels than model parameters but the differences are often too small to indicate unequivocally which model is best, especially when there are the usual sort of experimental error effects amongst the data. Examples of similar effects with these and other models are given in Colwell (1983) with functions estimated from data of experiments with 5 and 10 treatment levels.

The nonsense inference suggested by such effects of model is that fertilizer requirements are largely determined by choice of model. Thus

Table 5.1 Replicate means of example experimental data giving the estimated yield functions (5.1) to (5.4) for the relationship between crop yield Y and nutrient application rate P, illustrated in Fig. 5.1.

Y kg/ha	P kg/ha
600	0
2350	25
2670	81

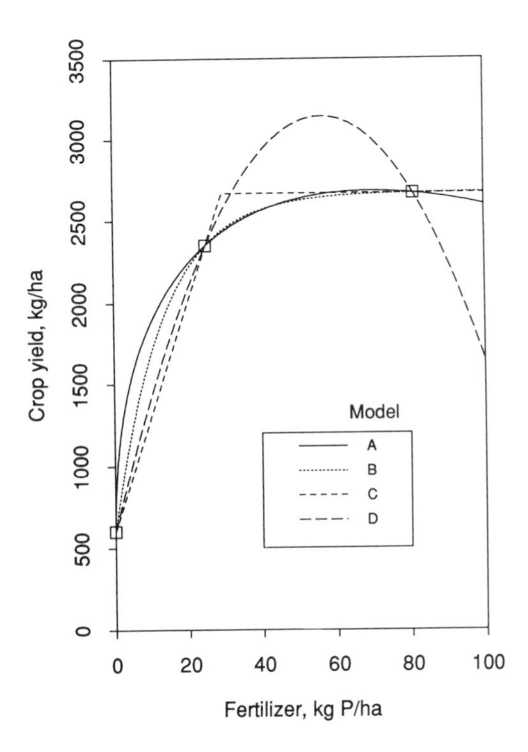

Fig. 5.1 Regressions estimated with the four alternative models

$$\text{A:}\quad Y = b_0 + b_1 N^{.5} + b_2 N$$

$$\text{B:}\quad Y = b_0 + b_1 \exp(b_2 N)$$

$$\text{C:}\quad Y = b_0 + b_1 \left(N - b_2 - |N - b_2|\right)$$

$$\text{D:}\quad Y = b_0 + b_1 N + b_2 N^2$$

from the data in Table 5.1. The □ indicate mean data values.

Table 5.2 Alternative models for single nutrient yield functions and optimal rate formulas derived from $dY/dN = E_n$.

Model	$Y = f(N)$	Optimal rate
A. Square root quadratic	$Y = b_0 + b_1 N^{.5} + b_2 N$	$N = \left(\dfrac{0.5b_1}{E_n - b_2}\right)^2$
B. Mitscherlich	$Y = b_0 + b_1 \exp(b_2 N)$	$N = \dfrac{1}{b_2}\log\left(\dfrac{E_n}{b_1 b_2}\right)$
C. "Broken Stick"	$Y = b_0 + b_1(N - b_2 - \lvert N - b_2 \rvert)$	$\left.\begin{array}{l} N = b_2 \text{ if } 2b_1 \geq E_n \\ N = 0 \text{ if } 2b_1 < E_n \end{array}\right\}$
D. Quadratic	$Y = b_0 + b_1 N + b_2 N^2$	$N = \dfrac{E_n - b_1}{2b_2}$

for example fertilizer requirements are usually greater when calculated from functions with the quadratic model form than when calculated from functions with the square root form (38.5 kg P/ha compared with 18.1 kg P/ha in the above examples). Although this is obviously nonsense it is also alarming because each of the example models in Table 5.2 has been used and advocated for the routine estimation of fertilizer requirements by respected scientists and statisticians. It would be interesting to know in this respect the extent to which choice of model for research on fertilizer requirements has determined the rates of fertilizer application that are used by farmers in different parts of the world.

The problem arises, presumably, because of undue reliance on standard statistical tests of significance and tests for goodness of fit with regressions estimated from data and a lack of consideration of the effects of model on estimates of interpolated values for the relationship as described below. Standard statistical tests give no indication of errors of interpolation (or extrapolation) because regression estimates of functions with a particular model form are obtained mathematically by calculating the regression equation that gives a best fit of the equation to the data in the sense that the sum of squares of deviations are data only for relatively few treatment levels, as is common with fertilizer experiments, regressions tend to fit the data closely so that sums of squares for deviations are relatively small and similar for different models. The above examples are extreme in this respect in that because there are only three treatment levels, the alternative models, each with three parameters, give identical fits to the data and consequently identical sums of squares of deviations, tests of significance, R^2 values and so on. Even however when there are 10 fertilizer treatment levels, differences in these values for different models may be relatively small although the forms of the estimated relationships may differ greatly (Colwell 1983). Clearly the choice of model for the representation of the fertilizer-yield relationship should not be based solely on standard statistical tests of significance or goodness of fit but also on considerations of the effect of model on the form of estimated relationships, or more specifically on the effects of model on estimates by interpolation and extrapolation.

This chapter is devoted to the choice of models for functions to represent the effects of fertilizer rate on yield. Bases for the choice are, (i) mathematical features of the models and notions concerning the *true* form of the relationship and (ii) computational convenience for the estimation of functions from experimental data, the calculation of optimal rates, the establishment of relationships between yield functions and site variables in a variable region, and for corresponding estimates with model extensions for multinutrient functions. These considerations, particularly those of computational convenience, led to the choice of square root quadratic models for the procedures described in this book.

Corresponding considerations could be made of modelling for the effects of location within experiment sites and of alternatives to the use of the orthogonal polynomials L_b, Q_b, etc. as dummy variables for block effects (chapter 2). Since however convenient designs are available for fertilizer experiments with relatively small blocks and since in general estimates are required for experiment sites as a whole rather than for particular locations within the sites, the representation of location effects in this way is generally sufficient.

5.2 Theoretical models

In the physical sciences, important and reliable models for relationships between variables have often been indicated by scientific theory and this may suggest that the problem in selecting a model for the fertilizer-yield relationship can be resolved similarly, by deriving a theoretical model. It is important to recognize however that theoretical models are only as good as the theory on which they are based and only for the range of conditions for which the theory is valid. For example the classical growth model for increase in population size with time

$$Y = b\exp(cT) \tag{5.5}$$

where Y is population size and T is time, is derived on the highly plausible basis that the rate of reproduction by the members of a population is constant for constant growing conditions. For this theoretical situation the rate of increase of the size of a population is simply proportional to its size as represented by the differential equation

$$\frac{dY}{dT} = cY \tag{5.6}$$

where c is a constant, and the growth model follows from this differential equation. Similarly the famous model of Mitscherlich (1909) for the fertilizer-yield relationship

$$Y = A - b\exp(-cN) \tag{5.7}$$

can be derived on the theoretical basis that the rate of crop yield response to nutrient application rate is proportional to the yield deficit from an attainable maximum A as represented by the differential equation

$$\frac{dY}{dN} = c(A - Y) \tag{5.8}$$

There is however an important difference. In this case the "theory" is neither based on convincing scientific evidence nor an acceptable premise but rather is suggested by the model itself, giving a circular argument. Experimental data (Colwell 1985a) may show in fact that rate of response dY/dN often has little relationship to the yield deficit $(A - Y)$.

The important point to remember is that there is no independent or reliable theory to indicate a model form for the fertilizer-yield relationship and given the complexity of the processes that determine plant growth and response to fertilizer, it is doubtful that there ever will be. Accordingly and whether we like it or not, models must be chosen empirically, on the subjective basis of observation and experience. The unfortunate feature of this necessity is that the *subjective basis* is often very subjective since the *observation and experience* is often not as extensive or critical as may be claimed by proponents of particular models. This is evidenced by the vastly different array of empirical models that have been chosen on this basis as illustrated with the four examples in Fig. 5.1.

5.3 Empirical models

Bases for a choice

In scientific investigations decisions are based on observation so that the choice of models for the scientific study of fertilizer-yield relationships should be based on the form of the relationship indicated by data for the effects of fertilizer rate on yield. There are however some rather obvious practical difficulties:

1. If statistical procedures are to be used to discriminate between models using statistical tests for goodness-of-fit and if models are to be chosen that are least likely to produce estimates affected by serious errors of interpolation or extrapolation, then data are required from experiments with many and widely ranging treatment rates. Thus an inspection of the graphs in Fig. 5.1 suggests that if data were obtained for two additional intermediate rates, say for 15 and 60 kg P/ha, data deviations from the estimated regressions, as used for statistical tests, would indicate the relative magnitudes of errors of interpolation for the alternative models. Similarly errors of extrapolation would be indicated by data for higher rates, say for 100 or 150 kg P/ha. Although data from additional well placed treatment rates would thus undoubtedly indicate the relative merits of alternative functions in Fig. 5.1, such ideal treatment rates for experiments usually cannot be chosen without the benefit of hindsight so that more than six or seven treatment rates might be required to indicate the relative merits of the alternative models given that growing conditions for future experiments may differ appreciably from those that produced the data for this example.
2. If experimental data are to be used to choose a model that can be used for functions representing the relationship under the variety of growing conditions that exist in an agricultural region, then data must be obtained to represent this variety of conditions. Given the need also for many treatment levels, particularly if the data are to cover interaction effects from application of two or more different nutrients, many large experiments at many sites and in many different kinds of growing season become necessary.
3. The form of the relationship may vary with growing conditions such as the seasonal weather, soil nutrient levels and interaction effects from applications of other nutrients, and with the fertilizer nutrient,

so that different model forms may be indicated by different sets of data.

4. The differences between estimated functions for alternative models are often small relative to error effects so that experiments need to be of high quality if they are to produce data that will discriminate accurately between alternative models. When the alternative models produce similar estimates of the relationship, at least at the application rates of the data from which they are estimated, normal levels of experimental error can make the discrimination between models impossible on a critical scientific and statistical basis.

The data requirements for a definitive choice of models indicated by these difficulties are virtually impossible to satisfy because of the time and expense involved, except perhaps for some special situations. Consequently, in practice, models must be selected mainly on the basis of notions concerning the actual or true form of the relationship. Often these are simply those suggested by a traditional model that has been used in the past, leading to circular arguments in favour of that model. Given this situation, the notions and reasons responsible for the choice of a particular model should at least be indicated by the scientist to allow critical evaluation. Accordingly the notions and reasons that have led to the choice of the square root quadratic model as the basis for procedures described in this book are listed below.

Notional features

Desirable features for models for the fertilizer-yield relationship are suggested by comparison of the features of the functions that are estimated with alternative models for a particular set of data and of the features of functions for each model, for the range of the types of relationship that the functions may be required to represent. This is illustrated above with the graphs for the four alternative models in Fig. 5.1 estimated from the data in Table 5.1 and similarly by the graphs in Figs 5.2 to 5.4 calculated from the sets of artificial data in Table 5.3 chosen to represent the range of types of relationship that might be expected in a variable region. Thus the data set Y_1 has been chosen to produce the types of function that are estimated with ideal experiments where the experimental data clearly define a maximum yield. Similarly the set Y_3 has been chosen to produce, in contrast, functions for a nearly straight line type of relationship such as are indicated when fertilizer treatment rates do not extend to a sufficiently high level to define a

Table 5.3 Artificial data for a range of types of yield response to fertilizer applications.

Fertilizer	Yield kg/ha		
kg P/ha	Y_1	Y_2	Y_3
0	0	1000	2000
25	2500	2500	2500
81	2800	2800	2800

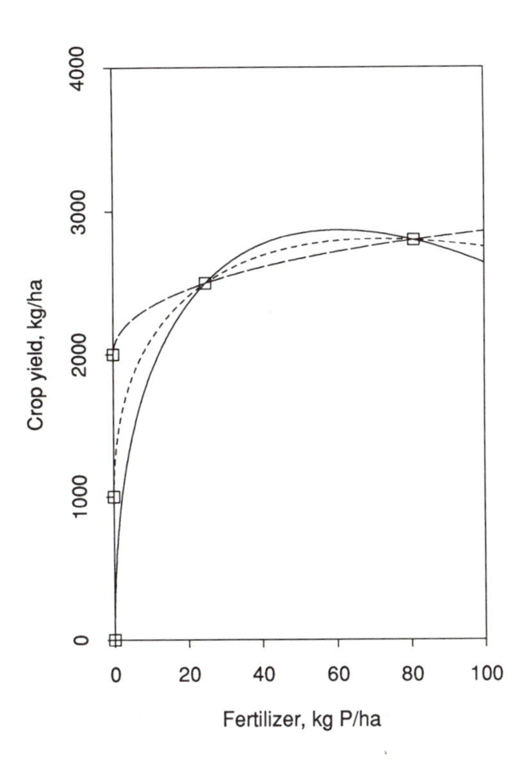

Fig. 5.2 Yield functions for the Y_1, Y_2 and Y_3 data of Table 5.3 produced by the square root quadratic model.

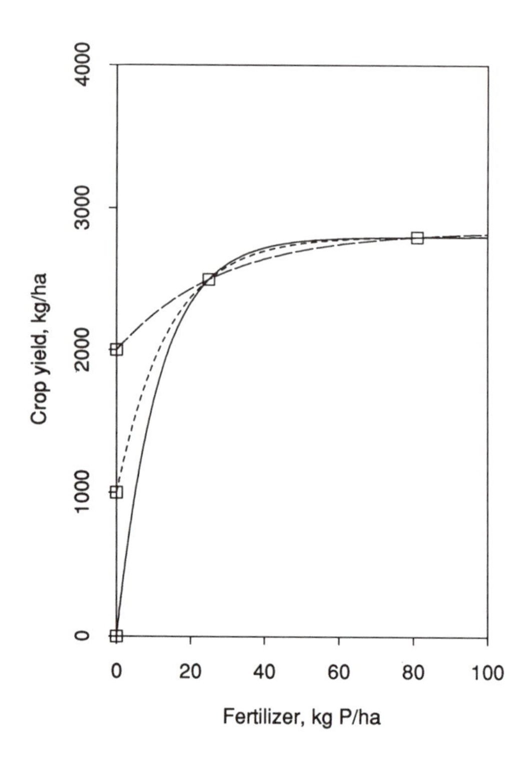

Fig. 5.3 As for Fig. 5.2, with the Mitscherlich model.

maximum yield. Finally the Y_2 set has been chosen to produce functions with an intermediate form between these extremes.

Functions calculated from these contrasting data sets for the alternative model forms in Table 5.2 are illustrated with graphs in Fig. 5.2 for the square root quadratic, Fig. 5.3 for the Mitscherlich and Fig. 5.4 for the quadratic model. Corresponding graphs for the "broken-stick" model are not given since for this model all functions have essentially the same form as curve C in Fig. 5.1, differing only in the slope and length of the first portion of the "stick". Corresponding illustrations for the range of functions likely to be estimated from experimental data may be obtained by varying the scale of the graphs and the position of their intersections on the Y axis but all will show the same contrasting features illustrated in these figures. The graphs

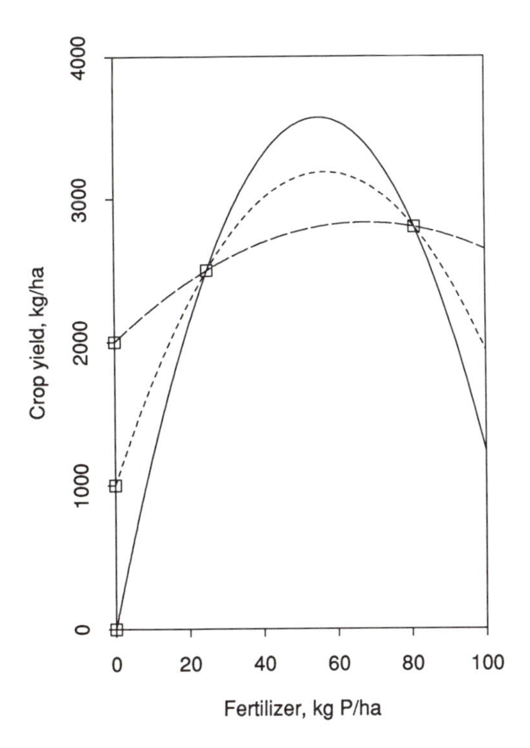

Fig. 5.4 As for Fig. 5.2, with the quadratic model.

indicate the nature of interpolation and extrapolation that are characteristic of the models and thus provide bases for choice against notions of the true forms of the relationships. The validity of the choice based on the comparisons of these graphs depends then on the validity of the notions which are now listed to allow critical evaluation.

1. *The relationship should have a continuous diminishing form such that the rate of yield response to fertilizer application rate, dY/dN, decreases progressively with increase in application rate towards that producing maximum yield where dY/dN = 0.*

The square root quadratic, Mitscherlich and quadratic functions accord with this notion but the broken-stick model (curve C in Fig. 5.1) does not. With it the rate of response $\mathrm{d}Y/\mathrm{d}N$ is constant up to the "break point" where it suddenly changes to zero and where consequently the optimal rate suddenly changes from b_2 to zero if $2b_1 \geq E_n$ (Table 5.2).

2. *Sufficiently high fertilizer application rates should produce near maximum yields for a wide range of rates and very high rates should depress yields.*

The square root quadratic model alone accords with this notion (Fig. 5.2). The Mitscherlich function (Fig. 5.3) is unsatisfactory in this respect since mathematically it can only define an asymptotic maximum for an infinitely high fertilizer rate. This however may not represent a serious defect for this model because the yield trend towards an asymptotic maximum can approximate to a region of near maximum yield and often there is no need to represent the yield depression from high application rates. The broken-stick function is satisfactory to the extent that it defines a broad range of constant maximum yield but it cannot represent a yield depression for very high rates. The quadratic model, Fig. 5.4, is clearly unsatisfactory however, because mathematically it must produce maximum curvature (d^2Y/dN^2) in the vicinity of the maximum rather than a broad plateau of near maximum yields. This mathematical feature of quadratics can produce unlikely kinds of maxima in functions as with the pronounced maxima in Fig. 5.4, even when the data from which they are estimated indicate continuing positive responses over the range of the data application rates.

3. *The rate of response dY/dN should not be very large for any fertilizer rate.*

The square root quadratic is unsatisfactory with respect to this notion with very low application rates because the rate of response becomes very large as the fertilizer rate approaches zero ($dY/dN \rightarrow \infty$ as $N \rightarrow 0$). This feature will usually not be important however because there is generally little interest in the form of the relationship for very low rates, that is for near zero application rates.

Evaluations with profit functions

Since optimal fertilizer rates are determined by the relationship between fertilizer rate and economic return, it may seem helpful to inspect the effect of model on the form of profit functions as illustrated in Fig. 5.5 with graphs for profit functions derived from the yield functions of Fig. 5.1 by the profit equation described in chapter 3

$$P = VY - C_pP(1+R)^t - Q$$

with the example economic values V = \$0.08, C_p = \$1.92, R = 0.02, t = 1 and neglecting Q for fixed costs.

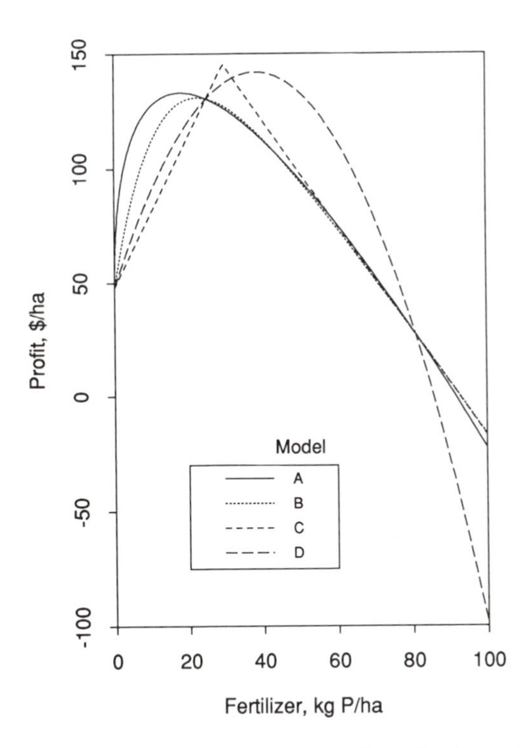

Fig. 5.5 Profit functions corresponding to the yield functions A to D of Table 5.2, illustrated in Fig. 5.1.

The most obvious effect of model on profit functions is on the position of the profit maximum for the optimal fertilizer rate. Thus for the example the optimal rates 18.1, 22.6, 29.6 and 38.5 kg P/ha, increase in the model sequence: square root quadratic < Mitscherlich < broken-stick < quadratic. Such sequences vary with the data from which functions are estimated and to a lesser extent with economic variables but in general, optimal rates calculated from square root quadratic and Mitscherlich models are appreciably lower than those calculated from quadratics as in this example. Apart however from emphasizing this effect of model on optimal rate and the consequent importance of using an appropriate model, the profit functions do not show the contrasting features of different models as clearly as the yield functions.

5.4 Computational convenience

Many elegant and ingenious mathematical models are available for the representation of relationships in agricultural science (Mead and Pike 1975, Sparrow 1979, France and Thornley 1984, Mead 1988) and undoubtedly many of these could be adapted to provide very satisfactory representations of the fertilizer-yield relationship. There is however a major obstacle to the use of most of them for the representation of the yield-fertilizer rate relationship, namely the complexity of the computations that their use entails with respect to:

1. the estimation of functions from experimental data, especially for multinutrient relationships,
2. the calculation of optimal rates,
3. analyses of variance,
4. the generalization of relationships in the form of *general soil fertility models* (chapters 8 and 9) that represent the effects of site variables on the relationship in a variable region, and in particular
5. the prediction of functions for sites in a region from values for site variables.

Although computations have become much more practicable with the development of modern computing facilities, they can nevertheless still present substantial problems so that computational convenience is still an important consideration. Consequently the experimenter must compromise between that which is possible or desirable in theory and that which is practicable, by assessing the significance of the features of alternative models and the labour and anguish that their use will entail. It is important to remember in this respect that all models are essentially empirical as indicated above and that when models give similar representations of the relationship, preferences in favour of a particular model form will usually depend on notions about the true nature of the relationship rather than convincing scientific evidence. Thus when a choice is to be made between two alternative models, the model requiring simpler computations will be preferred unless the differences are considered sufficiently significant to warrant the more complex computations. For example differences between regression estimates of functions for the models $Y = b_0 + b_1 N^{.5} + b_2 N$ and $Y = b_0 + b_1 \exp(b_2 N)$ are usually negligible except for very high or very low values of N, as illustrated in Fig. 5.1, so that the $Y = b_0 + b_1 N^{.5} + b_2 N$ model is preferred because its use entails simpler computations, as described below. Similarly although the "exponential plus linear model" $Y = b_0 + b_1 r^N + b_2 N$ where r is a "predetermined constant" (Sylvester-Bradley *et al.* 1984) and "inverse polynomial" models such as $\frac{N}{Y} = b_0 + b_1N + b_2N^2$ or $\frac{1}{Y} = \frac{1}{b_0} + \frac{b_1}{b_2 + N}$ (Balmukand 1928, Nelder 1966) have attractive features, the simpler model is preferred because of its computational convenience.

Multinutrient models

Multinutrient models can be derived simply by combining single nutrient models, assuming that each of the nutrients has the same kinds of effect on yield. Since however the magnitudes of the effects from the application of each nutrient are commonly affected by the rate of application of the others, provision must also be made for the representation of interaction effects. Thus if the individual effects of the nutrients N, P, K, ... can be represented by the single nutrient models $Y = f_n(N)$, $Y = f_p(P)$, $Y = f_k(K)$, ... , the form for multinutrient models can be derived in general terms by their combination with provision for interaction effects: for the two nutrient functions $Y = f_n(N)$ and $Y = f_p(P)$ by

$$Y = f_n(N) + f_p(P) + f_n(N)\,f_p(P) \tag{5.9}$$

for three nutrient functions $Y = f_n(N)$, $Y = f_p(P)$ and $Y = f_k(K)$ by

$$\begin{aligned} Y = {} & f_n(N) + f_p(P) + f_k(K) + f_n(N)\,f_p(P) \\ & + f_n(N)\,f_k(K) + f_p(P)\,f_k(K) \end{aligned} \tag{5.10}$$

and so on for four and more nutrient models. For example, given two single nutrient models $Y = b_0 + b_1N^{\cdot 5} + b_2N$ and $Y = b_0 + b_1P^{\cdot 5} + b_2P$, the mathematical extension to a two nutrient model corresponding to (5.9) would be

$$\begin{aligned} Y = {} & b_0 + b_1N^{\cdot 5} + b_2P^{\cdot 5} + b_3N + b_4P + b_5(NP)^{\cdot 5} \\ & + b_6N^{\cdot 5}P + b_7NP^{\cdot 5} + b_8NP \end{aligned} \tag{5.11}$$

Models derived in this way are however mostly unnecessarily large and complex, particularly when interaction effects are relatively small, so that they can be simplified by deleting terms that are generally found to make non-significant contributions to estimates of the relationship, as shown with analyses of variance. Thus studies with the data from about 300 nitrogen and phosphorus fertilizer experiments with wheat in Australia (Colwell 1977, 1979) showed that the contributions of the higher order interaction terms $N^{\cdot 5}P$, $NP^{\cdot 5}$ and NP to functions estimated for the above example model were generally non-significant in the presence of the $(NP)^{.5}$ term so that they can be omitted to obtain the simpler and more convenient model

$$Y = b_0 + b_1 N^{.5} + b_2 P^{.5} + b_3 (NP)^{.5} + b_4 N + b_5 P \quad (5.12)$$

Similarly the relatively simple three and four nutrient models

$$Y = b_0 + b_1 N^{.5} + b_2 P^{.5} + b_3 K^{.5} + b_4 (NP)^{.5} + b_5 (NK)^{.5} + b_6 (PK)^{.5} + b_7 N + b_8 P + b_9 K \quad (5.13)$$

and

$$Y = b_0 + b_1 N^{.5} + b_2 P^{.5} + b_3 K^{.5} + b_4 S^{.5} + b_5 (NP)^{.5} + b_6 (NK)^{.5} + b_7 (NS)^{.5} + b_8 (PK)^{.5} + b_9 (PS)^{.5} + b_{10} (KS)^{.5} + b_{11} N + b_{12} P + b_{13} K + b_{14} S \quad (5.14)$$

can be derived on the basis that the higher order interaction terms $N^{.5}P$, NP, $N^{.5}P^{.5}K^{.5}$, $NP^{.5}K^{.5}$, etc. are unnecessary in the presence of terms for linear × linear interaction effects (square root scale).

Multinutrient models derived in this way with non-polynomial models mostly require much more complex computations for the estimation of functions from experimental data compared with those for polynomials, especially when the models are "non-linear" in the statistical sense[1].

Estimation of functions

When a model has been selected for the fertilizer-yield relationship, functions with that form have to be estimated, usually on a routine scale, as when data have been obtained from a series of fertilizer experiments. If the model has a statistical *linear*[1] form, estimates of the functions can be computed directly and simply by the standard least squares regression procedure but if the model has a *non-linear* form, estimates must be computed using more difficult procedures, requiring iteration and successive approximation. Thus functions of the square root quadratic and quadratic model forms and their extensions for multinutrient models can be estimated conveniently and directly by the standard regression procedure whereas those of the Mitscherlich, broken-stick and inverse polynomial forms cannot. The "exponential

[1] The terms *linear* and *non-linear* in this usage refer to the model parameters and not the variables. Thus polynomial models are linear whereas models like the Mitscherlich are non-linear even though all can represent curved relationships.

plus linear" model is special in this regard being linear and simple to estimate if there is a predetermined value for the variable r (Sylvester-Bradley *et al.* 1984 nominate $r = 0.99$) but non-linear and requiring an iterative estimation procedure if r also has to be estimated. In addition the estimates of the parameters for linear functions are themselves linear functions of the data from which they have been estimated so that it is simple to determine their means, variances and covariances and hence to compute standard errors of expressions derived from the regression coefficients as described in chapter 4. The advantages of linear models in these respects become very substantial with multinutrient functions.

Calculation of optimal rates

The calculation of optimal rates from single nutrient functions by solution of the equation $dY/dN = E_n$ is relatively simple with most model forms (Tables 3.1 and 5.2) but become much more complex with multinutrient functions as seen in the formulas (3.10) and (3.11) for the two and three nutrient extensions of the square root quadratic model. Nevertheless the calculations for multinutrient quadratic models are direct, requiring only the use of standard linear algebraic procedures for the solution of simultaneous equations. But for multinutrient versions of non-linear models, direct solutions are mostly not possible and require the use of inconvenient successive approximation procedures.

Yield variables for relationships with site variables

The establishment of relationships between yield functions and site variables that can be used to predict functions for sites in a variable region is very much simpler if the functions can be represented by variables corresponding to orthogonal components of analyses of variance as described in following chapters. This can be done relatively simply with polynomial functions by converting them to an orthogonal form but corresponding procedures are not available for most other models. This is a major reason for the use of polynomial models for the fertilizer-yield relationship for studies on soil fertility relationships or soil test calibrations for the estimation of fertilizer requirements as described in chapters 8 and 9. In particular this is a reason for not using the popular Mitscherlich model for studies on soil fertility relationships, even though when it is rearranged in the form $Y = a[1 - \exp\{c(N + b)\}]$, biological meanings can be associated with each of its parameters, i.e. a = maximum attainable yield, b = soil nutrient content and c = a nutrient efficiency or availability factor.

5.5 Choice of the square root quadratic model

The above considerations of alternative models and their extensions for multinutrient relationships have led to the use of the square root quadratic model for the representation of the yield-fertilizer rate relationship in this book. Although it is not ideal, as indicated by the "bad features" listed below, it gives a good representation of the relationship for the more important range of application rates in the vicinity of the optimal rate and can be used for the development of general soil fertility models. It also usually gives remarkably close representations of the relationship to those given by the Mitscherlich model, as with the example in Fig. 5.1, or by the "exponential plus linear" model, as can be seen by comparing accurate computer drawn graphs for the alternative functions estimated from the same sets of data. In conclusion, then, the *good* and *bad* features of the square root quadratic model are summarized to indicate the reason for the general use of this model for the procedures described in this book.

Good features

1. The square root quadratic model gives a good representation of important aspects of the notional form of the relationship, namely a smooth diminishing response with increase in fertilizer rate towards a broad plateau of near maximum yields. Thus errors with estimates by interpolation seem unlikely to be serious within the important range of fertilizer application rates in the vicinity of the optimal rate and that for maximum yield.
2. It is readily extended for the representation of multinutrient relationships with provision for interaction effects.
3. It is computationally convenient for (i) the estimation of functions from experimental data by the standard regression procedure, (ii) the calculation of standard errors for estimates made with the functions and (iii) the calculation of optimal fertilizer rates.
4. Estimates of optimal fertilizer rates are generally lower than those obtained from functions for other models. Given the environmental problems that can be caused by excessive applications of fertilizers, the use of the generally more conservative estimates obtained with functions of this form is preferred, at least until convincing evidence is obtained to indicate the need for some alternative model.

5. In addition, as will be described in following chapters, polynomial regressions can be rearranged in an orthogonal form suitable for the study of relationships of aspects of soil fertility with site variables in a variable region and for statistical predictions of yields and fertilizer requirements from soil analyses and other site variables.

Bad features

1. The model is not ideal in that estimated functions give high rates of response to fertilizer applications for very low, near zero rates of application. Ordinarily however there is little interest in the yield produced by near zero fertilizer application rates so that this bad feature is not important. It may simply be noted then as a feature that precludes the use of functions with this form for estimating yield response to very low fertilizer rates.
2. As with most regressions, estimates involving extrapolation are likely to be unreliable. The graphs in Figs 5.1 and 5.2 suggest that such errors may be relatively small for small extrapolations to higher rates but that extrapolations to lower rates when estimates have been obtained from data that do not include the zero rate will be particularly unreliable.

Chapter 6

Analysis of Polynomial Functions

6.1 Introduction

Polynomial models such as $Y = b_0 + b_1X$, $Y = b_0 + b_1X + b_2X^2$, $Y = b_0 + b_1X + b_2W + b_3XW + b_4X^2 + b_5W^2$, etc. are convenient and popular mathematical forms for empirical regression equations representing relationships between a dependent variable Y and regressor variables X, W, etc. For convenience the models can be represented as members of an hierarchical series as in Table 6.1, commencing with the polynomial of degree 0 for a nil relationship where there is no regressor variable and the "regression" is simply the equation of a constant $Y = b_0$ estimated by a data mean $b_0 = \overline{Y}$, followed by polynomials of degree 1 for linear or straight line relationships, degree 2 for quadratic types of curved relationship, degree 3 for cubic types of curved relationship and so on to higher degrees for more complex types of relationship. Furthermore, the hierarchical series can be extended to represent relationships with two or more variables as with X, W, V in Table 6.1 and to other types of curved relationship by converting the variables to other mathematical scales, as by substituting $N^{.5} = X$, $P^{.5} = V$, etc. to obtain the corresponding series of *polynomials on the square root scale* in Table 6.2. In summary, such polynomial models are convenient for the representation of the fertilizer-yield relationship because:

1. They have a statistically linear form so that regressions can be estimated simply by the usual least squares regression procedure.
2. They can be used as models for multivariate relationships with provision for interaction effects.

3. Values for maxima or minima can be calculated by relatively simple procedures.
4. Polynomial regressions can be transformed algebraically into an orthogonal form such that the regressor variables are polynomial expressions that are orthogonal with respect to each other for particular values of the regressor variables.
5. Regressions can be separated into components corresponding to the orthogonal components of a detailed analysis of variance.
6. In particular, the series in Table 6.2, termed simply *square root polynomials,* give good representations of the fertilizer-yield relationship.

This chapter describes statistical procedures with polynomial regressions that provide the bases for procedures to be described in following chapters for (i) detailed analyses of variance of experimental data, (ii) the design of experiments suitable for providing data for the estimation of polynomial regression relationships and (iii) the establishment of predictive relationships between the results of fertilizer experiments and site variables. Although the mathematics associated with these procedures may appear rather awesome to the non-mathematician, a general understanding of their nature will provide an appreciation of the traps that exist for the unwary in the seemingly obvious application of some standard statistical procedures. Sections marked with # may be skipped by readers not interested in the mathematical bases for the procedures.

6.2 Terminology

Degree

The *degree* of a polynomial term is defined by the sum of its exponents and the degree of a polynomial function is determined by the highest degree of any term it contains. For example the terms XY^2 , XYZ and X^3 are all of degree 3 and if these are the highest degree of any term in a polynomial the polynomial is described as being of degree 3. The term *degree zero* is also used to describe a constant since $X^0 = 1$ so that a constant c may be regarded as being the polynomial term cX^0. Thus equations $Y = b_0$ and $Y = \overline{Y}$ may be described as being polynomials of zero degree.

Polynomials in which the term of highest degree is X, X^2 and X^3 are also described as being linear, quadratic and cubic functions.

Table 6.1 Polynomial functions on the natural scale.

Polynomial on the natural scale	Degree	Function name
$Y = b_0$ or $Y = \bar{Y}$	0	Constant
$Y = b_0 + b_1X$	1	Linear
$Y = b_0 + b_1X + b_2X^2$	2	Quadratic
$Y = b_0 + b_1X + b_2X^2 + b_3X^3$	3	Cubic
$Y = b_0 + b_1X + b_2X^2 + b_3X^3 + \cdots + b_nX^n$	n	*n* th degree
$Y = b_0 + b_1X + b_2W + b_3XW + b_4X^2 + b_5W^2$	2	Quadratic
$Y = b_0 + b_1X + b_2W + b_3V + b_4XW + b_5XV$ $+ b_6WV + b_7X^2 + b_8W^2 + b_9V^2$	2	Quadratic

Table 6.2 Corresponding polynomial functions to those in Table 6.1, on the square root scale.

Polynomial on the square root scale	Degree	Function name
$Y = b_0$ or $Y = \bar{Y}$	0	Constant
$Y = b_0 + b_1N^{.5}$	1	Linear
$Y = b_0 + b_1N^{.5} + b_2N$	2	Quadratic
$Y = b_0 + b_1N^{.5} + b_2N + b_3N^{1.5}$	3	Cubic
$Y = b_0 + b_1N^{.5} + b_2N + b_3N^{1.5} + \cdots + b_nN^{0.5n}$	n	*n* th degree
$Y = b_0 + b_1N^{.5} + b_2P^{.5} + b_3(NP)^{.5} + b_4N + b_5P$	2	Quadratic
$Y = b_0 + b_1N^{.5} + b_2P^{.5} + b_3K^{.5} + b_4(NP)^{.5}$ $+ b_5(NK)^{.5} + b_6(PK)^{.5} + b_7N + b_8P + b_9K$	2	Quadratic

Scale

The polynomial notation and procedures can be readily extended for use with other types of relationships by the simple substitution $X = N^S$ to obtain polynomials for *N on the S scale* in place of the ordinary polynomials for *X on the natural scale* as in Table 6.1. In particular the substitution $X = N^{\cdot 5}$, $W = P^{\cdot 5}$, etc. gives polynomials *on the square root scale* as in Table 6.2, the quadratics on the square root scale, as used for fertilizer-yield functions, being termed simply *square root quadratics*.

Orthogonal

The sets of n numbers X_i and W_i, $i = 1, 2, \ldots, n$ are said to be orthogonal with respect to each other if the sum of their products is zero

$$\sum_{i=1}^{n}\sum_{j=1}^{n} X_i W_j = 0 \quad \text{if } i \neq j$$

For example the four values for the variables Z, L, Q and C in Table 6.3 are all orthogonal with respect to each other because

$$\sum_{i=1}^{4} Z_i L_i = \sum_{i=1}^{4} Z_i Q_i = \sum_{i=1}^{4} Z_i C_i = \sum_{i=1}^{4} L_i Q_i = \sum_{i=1}^{4} L_i C_i = \sum_{i=1}^{4} Q_i C_i = 0 \qquad (6.1)$$

It follows that the correlation between orthogonal variables is zero, $r = 0$, so that orthogonal variables may also be described as being independent of each other. Such sets of orthogonal values are useful if they can be used as data for regressor variables in regressions because their orthogonality has the mathematical consequence that the

Table 6.3 Orthogonal polynomial values for four equally spaced values of a variable.

Trend	Z	L	Q	C
	+1	−3	+1	−1
	+1	−1	−1	+3
	+1	+1	−1	−3
	+1	+3	+1	+1
Divisor	2	$\sqrt{20}$	2	$\sqrt{20}$

regression coefficients provide an independent measure of the effect of each orthogonal variable in the regression equation. It is for this reason that orthogonal values are used for dummy variables to represent block effects in the regression representations of experimental results, as with the values for L_b and Q_b in Table 2.2 and the regressions (2.4) and (2.5) for the introductory experiment. The fact that the coefficients of these variables provide independent measures of the respective block effects is seen by their constant values in alternative regression equations as in the regressions (2.4) and (2.5) and other example yield functions. For the same mathematical reason, variables that have orthogonal values for fertilizer treatment rates are used for analyses of variance and to represent treatment effects for studies on the relationships between the results of fertilizer experiments and site variables in variable regions as described in following chapters.

Sets of values for Z, L, Q and C, such as the example set in Table 6.3, represent zero, linear, quadratic and cubic polynomial trends or contrasts and they are consequently described as being *orthogonal trends* or *orthogonal contrasts* and their values as being values for *orthogonal polynomials.*

If orthogonal values for variables are adjusted so that their sums of squares are 1, as for example by dividing the values in Table 6.3 by the divisor at the bottom of each column so that

$$\sum_{i=1}^{4} Z_i^2 = \sum_{i=1}^{4} L_i^2 = \sum_{i=1}^{4} Q_i^2 = \sum_{i=1}^{4} C_i^2 = 1 \qquad (6.2)$$

they become mathematically very convenient for computing statistical analyses of variance of experimental data. Orthogonal polynomial values that have been adjusted in this way are described as being *normalized.* Thus the divisors in Table 6.3 are the square roots of the respective sums of squares of the values in each column.

6.3 Orthogonal polynomials

Polynomials of the variable X can be rearranged algebraically to replace the variables X^0, X^1, X^2, ... with polynomial expressions of X, often denoted by ξ_0, ξ_1, ξ_2, ... as in the statistical tables of Fisher and Yates (1963), which have values that are orthogonal with respect to each other for a nominated set of values of X. Thus for example the polynomial $Y = b_0 + b_1X + b_2X^2 + b_3X^3$ can be rearranged in the form $Y = p_0\xi_0 + p_1\xi_1 + p_2\xi_2 + p_3\xi_3$ where the ξ_0, ξ_1, ξ_2 and ξ_3 are polynomial expressions of X that are chosen so that their values are orthogonal with respect to each other for a nominated set of values of X. More conveniently for our purposes, the expressions ξ_0, ξ_1, ξ_2 and ξ_3 are denoted by the variables Z, L, Q and C since they are respectively polynomials of degree 0, 1, 2 and 3 that may be termed zero, linear, quadratic and cubic polynomials, the example $Y = p_0\xi_0 + p_1\xi_1 + p_2\xi_2 + p_3\xi_3$ becoming $Y = p_0Z + p_1L + p_2Q + p_3C$. The values for Z, L, Q and C in Table 6.3 are example values for such expressions calculated for four evenly spaced values of X, such as $X = 0, 1, 2, 3$ or 10, 30, 50, 70 and similar values can be calculated for other uneven spacings such as √0, √10, √25, √50 where the 0, 10, 25 and 50 might be nutrient treatment rates for a fertilizer experiment and the Z, L, Q and C values are required to estimate an orthogonal polynomial regression on the square root scale from experimental data.

The important points to note about orthogonal polynomials for our present purposes are:

1. Orthogonal polynomial regressions give identical representations of relationships to the corresponding ordinary polynomial regressions and merely differ in appearance because of the use of polynomial expressions as regressor variables in place of the simple X^0, X^1, X^2, ... of ordinary polynomials. For example an orthogonal polynomial regression $Y = p_0Z + p_1L + p_2Q + p_3C$ gives a mathematically identical estimation of a relationship to that obtained with an ordinary polynomial regression with the form $Y = b_0 + b_1X + b_2X^2 + b_3X^3$, although in a different mathematical form, when each is estimated from the same set of data.
2. Values for the coefficients b_0, b_1, b_2, ... of ordinary polynomial regressions can be calculated from the coefficients p_0, p_1, p_2 ... of orthogonal polynomial regressions and *vice versa*, by expanding the polynomial expressions that are used as regressor variables in the orthogonal regressions.
3. Mathematical calculations, as for estimates of yield or optimal rates from fertilizer-yield functions, are most conveniently made with the polynomial regressions in the non-orthogonal form.
4. Orthogonal polynomials can be obtained on the scale s simply by substituting $X = N^s$ in the expressions for the polynomial values. In particular the substitution $X = N^{.5}$ is used to obtain orthogonal polynomials on the square root scale.
5. Values for normalized orthogonal polynomials depend on the relative spacing of the values for the variable X and not on their magnitude. Thus the values X = 0, 1, 2, 3; X = –3, –1, 1, 3; X = 20,

40, 60, 80 all produce the same normalized orthogonal polynomial values for four evenly spaced values of a variable X.

6. Although in theory orthogonal polynomials of any degree can be derived, in practice the trends represented by polynomials of degree greater than 2 or 3 are likely to have little physical or biological meaning, at least for the representation of the fertilizer-yield relationship, so that normally there is no need to use polynomials of degree greater than 3.
7. Orthogonal polynomials for multivariate polynomials can be derived from orthogonal polynomials for single variables. For example the orthogonal polynomial

$$Y = p_0 Z + p_1 L_n + p_2 L_p + p_3 L_n L_p + p_4 Q_n + p_5 Q_p$$

corresponding to the yield function

$$Y = b_0 + b_1 N^{.5} + b_2 P^{.5} + b_3 (NP)^{.5} + b_4 N + b_5 P$$

can be derived from orthogonal polynomials of $N^{.5}$ and $P^{.5}$.
8. Orthogonal polynomial regressions are used in the following chapters to provide the mathematical basis for (a) obtaining orthogonal analyses of variance of fertilizer experiment data with regression equations, (b) selecting experimental designs and treatments for the estimation of fertilizer-yield functions, and (c) developing procedures for estimating predictive relationships between the results of fertilizer experiments and site variables in a variable region. The procedures thus developed do not require the actual use of orthogonal polynomials by the researcher, except as already noted, to provide values for dummy variables for block effects in fertilizer-yield functions.

#Derivation

The definitions of orthogonality and normality in equations (6.1) and (6.2) give the procedures for deriving polynomial expressions

$$Z_i = a_0$$

$$L_i = b_1(X_i - \overline{X})$$

$$Q_i = a_2 + b_2(X_i - \overline{X}) + c_2(X_i - \overline{X})^2$$

$$C_i = a_3 + b_3(X_i - \overline{X}) + c_3(X_i - \overline{X})^2 + d_3(X_i - \overline{X})^3 \qquad (6.3)$$

for the zero, linear, quadratic and cubic orthogonal trends Z_i, L_i, Q_i and C_i which are orthogonal for the n values X_i, i = 1, 2, ... , n of the variable X, follow from the definition of orthogonality and normality in equations (6.1) and (6.2). Thus if $\overline{X}$ is the mean of the X_i, values for a_0, b_1, ... d_3 are calculated sequentially commencing with $a_0 = \frac{1}{\sqrt{n}}$ so that $\sum_{i=1}^{n} Z_i^2 = 1$, then finding the value b_1 such that $\sum_{i=1}^{n} Z_i L_i = 0$ and $\sum_{i=1}^{n} L_i^2 = 1$, then the values a_2, b_2, c_2 such that $\sum_{i=1}^{n} Z_i L_i = \sum_{i=1}^{n} Z_i Q_i = \sum_{i=1}^{n} L_i Q_i = 0$ and $\sum_{i=1}^{n} Q_i^2 = 1$ and finally values for a_3, b_3, c_3 and d_3 such that $\sum_{i=1}^{n} Z_i C_i = \sum_{i=1}^{n} L_i C_i = \sum_{i=1}^{n} Q_i C_i = 0$ and $\sum_{i=1}^{n} C_i^2 = 1$. Corresponding equations can be derived for the computation of orthogonal polynomials of higher degree and for polynomials with two or more variables[1]. Using the abbreviated notation Σx, Σx^2, Σx^3, ... for the summations $\sum_{1=1}^{n}(X_i - \overline{X})$, $\sum_{1=1}^{n}(X_i - \overline{X})^2$, $\sum_{1=1}^{n}(X_i - \overline{X})^3$, ... values for the coefficients in (6.3) are obtained with the formulas:

$$a_0 = \frac{1}{\sqrt{n}} \tag{6.4}$$

$$b_1 = \frac{1}{(\Sigma x^2)^{\cdot 5}} \tag{6.5}$$

$$\left.\begin{aligned} a_2 &= -\frac{\Sigma x^2}{un}, \quad b_2 = -\frac{\Sigma x^3}{u\Sigma x^2}, \quad c_2 = \frac{1}{u} \\ \text{where } u &= \left(\Sigma x^4 - \frac{\left(\Sigma x^2\right)^2}{n} - \frac{\left(\Sigma x^3\right)^2}{\Sigma x^2} \right)^{\cdot 5} \end{aligned}\right\} \tag{6.6}$$

---

[1] A Fortran program for computing values for the constants of orthogonal polynomials and polynomial values for single and multifactor functions is given by Colwell (1978). A simple procedure is possible when values of the X_i are evenly spaced as described by Fisher (1937) and Snedecor and Cochran (1967).

$$a_3 = \frac{r}{v}\ ,\ b_3 = \frac{s}{v}\ ,\ c_3 = \frac{t}{v}\ ,\ d_3 = \frac{1}{v}$$

where

$$r = \frac{\left(\Sigma x^3\right)^3 + (\Sigma x^2)^2\,\Sigma x^5 - 2\,\Sigma x^2\,\Sigma x^3\,\Sigma x^4}{w}$$

$$s = \frac{\Sigma x^4\,(\Sigma x^2)^2 - \Sigma x^2\,(\Sigma x^3)^2 + n\Sigma x^3\,\Sigma x^5 - n(\Sigma x^4)^2}{w} \qquad (6.7)$$

$$t = \frac{(\Sigma x^2)^2\,\Sigma x^3 + \Sigma x^3\,\Sigma x^4 - n\Sigma x^2\,\Sigma x^5}{w}$$

$$v = \left(\sum_{i=1}^{n}\left\{r + s(X_i - \overline{X}) + t(X_i - \overline{X})^2 + (X_i - \overline{X})^3\right\}^2\right)^{.5}$$

$$w = n\Sigma x^2\,\Sigma x^4 - n(\Sigma x^3)^2 - (\Sigma x^2)^3$$

For the applications to be described, actual computations with these formulas are not necessary and they are given here only to indicate the bases for the conversion equations given below relating the coefficients of orthogonal polynomial regressions to those of corresponding non-orthogonal polynomial regressions, and for the general soil fertility models to be described in chapters 8 and 9.

#Component trends of polynomial regressions

When a polynomial regression such as

$$Y = b_0 + b_1X + b_2X^2 + b_3X^3$$

is written in the orthogonal form,

$$Y = p_0Z + p_1L + p_2Q + p_3C$$

the magnitudes of the coefficients p_0, p_1, p_2 and p_3 indicate the amount of the orthogonal zero, linear, quadratic and cubic trends represented by the variables Z, L, Q and C in the regression and the equations $Y = p_0Z$, $Y = p_1L$, $Y = p_2Q$ and $Y = p_3C$ provide an analysis of the regression into zero, linear, quadratic and cubic components. The nature of the trends thus separated is illustrated by the orthogonal polynomial values in Table 6.4, computed by the above equations for the evenly spaced values $X = 0$, 1, 2, 3 and the unevenly spaced values

Table 6.4 Orthogonal polynomial values for X = 0, 1, 2, 3 and X = 0, 1, 2, 4, from equations (6.8) and (6.9).

X	Z	L	Q	C
0	0.5	– 0.67082	0.5	– 0.22361
1	0.5	– 0.22361	– 0.5	0.67082
2	0.5	0.22361	– 0.5	– 0.67082
3	0.5	0.67082	0.5	0.22361
0	0.5	– 0.59161	0.56408	– 0.28604
1	0.5	– 0.25355	– 0.32233	0.76277
2	0.5	0.08452	– 0.65566	– 0.57208
4	0.5	0.76064	0.40291	0.09535

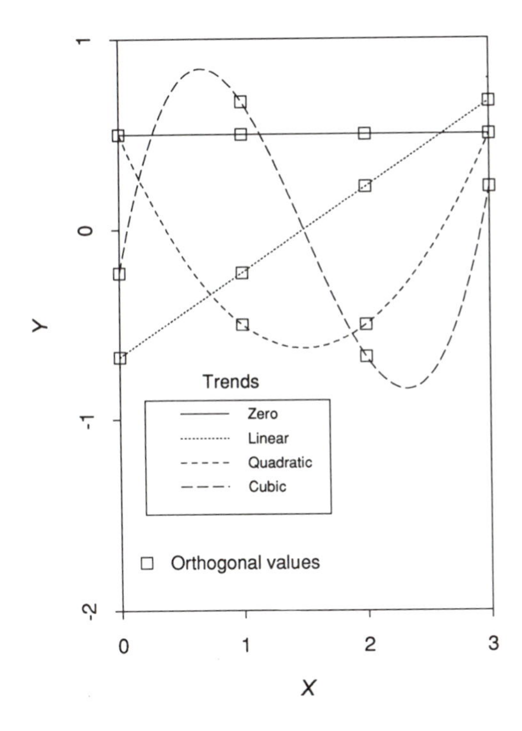

Fig. 6.1 Graphs of the trend equations (6.8) for orthogonal polynomials for the evenly spaced values X = 0, 1, 2, 3.

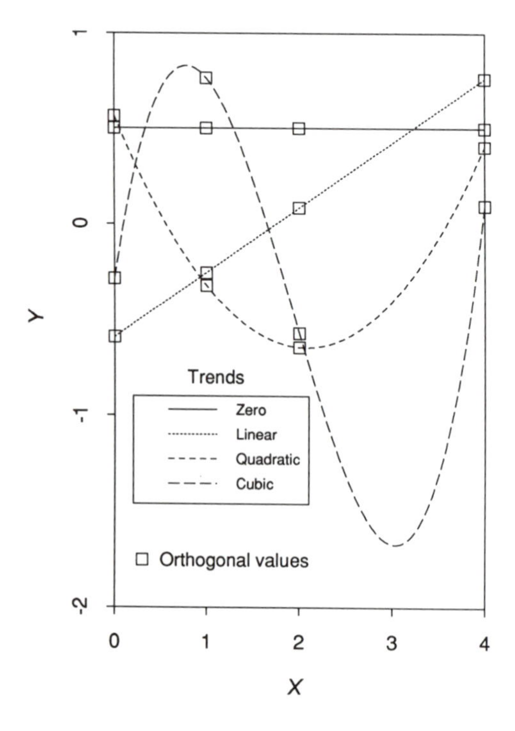

Fig. 6.2 Graphs of trend equations (6.9) for orthogonal polynomials for the unevenly spaced values $X = 0,1,2,4$.

$X = 0, 1, 2, 4$ and by corresponding graphs in Figs 6.1 and 6.2 for the trends defined by the respective equations corresponding to (6.3),

$$Y = Z = 0.5$$

$$Y = L = 0.44721(X - 1.5)$$

$$Y = Q = -0.625 + 0.5(X - 1.5)$$

$$Y = C = -1.528(X - 1.5) + 0.74536(X - 1.5)^2 \qquad (6.8)$$

and

$$Y = Z = 0.5$$

$$Y = L = 0.33806(X - 1.75)$$

$$Y = Q = -0.61696 - 0.18131(X - 1.75) + 0.28204(X - 1.75)^2$$

$$Y = C = -0.15829 - 1.63032(X - 1.75) - 0.20857(X - 1.75)^2 + 0.437(X - 1.75)^3 \quad (6.9)$$

Polynomial values and the form of the trends vary with the relative spacing of the X values as illustrated with these examples. The forms for the trends shown in Figs 6.1 and 6.2 are important because they indicate the nature of the interpolation between the data values of X in estimates from polynomial regressions. Thus Fig. 6.2 shows that interpolated values for cubic trends in regressions estimated from data with the relative spacings $X = 0, 1, 2, 4$ involve a large interpolation peak between $X = 2$ and 4 and hence indicates a potential for a serious interpolation error for estimates involving this trend in this range, that is from cubic regressions. Graphs for component polynomial trends thus serve to indicate desirable relative spacings for experiment treatment rates as will be described in Chapter 7.

#6.4 Regression analyses of variance with orthogonal polynomials

Given data for two variables X and Y, regression estimates of a polynomial relationship may be obtained with the regression expressed in either the non-orthogonal or the orthogonal polynomial form. Thus with the usual matrix notation, if Y data are represented by a vector $\mathbf{y}$ and the regressor variable data $1, X, X^2$ ... for a non-orthogonal polynomial are represented by the matrix $\mathbf{X}$ then the vector $\mathbf{b}$ for the coefficients in the regression

$$\mathbf{y} = \mathbf{Xb}$$

is obtained by

$$\mathbf{b} = (\mathbf{X}'\mathbf{X})^{-1}\mathbf{X}'\mathbf{y} \quad (6.10)$$

Similarly if the regressor variable data $\xi_0, \xi_1, \xi_2, \ldots$ for an orthogonal polynomial regression are represented by the matrix $\mathbf{E}$ then the vector $\mathbf{p}$ for the coefficients of the regression

$$\mathbf{y} = \mathbf{Ep}$$

is obtained by

$$\mathbf{p} = (\mathbf{E}'\mathbf{E})^{-1}\mathbf{E}'\mathbf{y} \tag{6.11}$$

In this latter case however, since the matrix **E** is orthogonal $\mathbf{E}'\mathbf{E} = \mathbf{I}$, where **I** is the identity matrix so that the vector **p** is obtained simply by

$$\mathbf{p} = \mathbf{E}'\mathbf{y} \tag{6.12}$$

Also because **E** is orthogonal, the squares of the trend coefficients p_0^2, p_1^2, $\cdots$ have the values of independent (orthogonal) components of the regression sum of squares. They consequently provide a simple means for obtaining a detailed regression analysis of variance as illustrated in Table 6.5 for the cubic regression

$$Y = b_0 + b_1X + b_2X^2 + b_3X^3 \quad \text{or} \quad Y = p_0Z + p_1L + p_2Q + p_3C$$

Table 6.5 The form of analyses of variance for cubic polynomials derived from trend coefficient values in the orthogonal regression $Y = p_0Z + p_1L + p_2Q + p_3C$.

Source	d.f.	Sum of squares
Mean or zero trend	1	p_0^2
Linear trend	1	p_1^2
Quadratic trend	1	p_2^2
Cubic trend	1	p_3^2
Regression	3	$p_1^2 + p_2^2 + p_3^2$
Residue	$n-4$	$\sum_{i=1}^{n} Y_i^2 - p_0^2 - p_1^2 - p_2^2 - p_3^2$
Total	$n-1$	$\sum_{i=1}^{n}\left(Y_i - \bar{Y}\right)^2 = \sum_{i=1}^{n} Y_i - p_0^2$

The identical detailed analysis of variance can be obtained from the data mean $\overline{Y}$ and analyses of variance for the non-orthogonal regressions

$$Y = a_0 + a_1X \tag{6.13}$$

$$Y = b_0 + b_1X + b_2X^2 \tag{6.14}$$

$$Y = c_0 + c_1X + c_2X^2 + c_3X^3 \tag{6.15}$$

Thus $p_0^2 = n\overline{Y}^2$

p_1^2 = the regression sum of squares for (6.13),

p_2^2 = the difference between the regression sums of squares for (6.13) and (6.14) and

p_3^2 = the difference between the regression sums of squares for (6.14) and (6.15).

Although this procedure involves more computations than simply squaring the orthogonal trend coefficients in the orthogonal regression, it is much more convenient given a modern computer and a program for a standard regression analysis of variance, since it obviates the need to compute orthogonal polynomial values for the orthogonal regressions. The mathematical nature of the orthogonal polynomials is simply used to indicate the equivalent non-orthogonal regression procedure.

6.5 Polynomials on the square root scale

The simple substitution $N^{\cdot 5} = X$ in the above polynomials gives polynomials *on the square root scale* and allows the use of the same computational procedures as those described above for polynomials on the natural scale. For example, trends on the square root scale corresponding to those illustrated in Fig. 6.1 for the natural scale are obtained by plotting the same graphs but substituting $N = X^2$ (because $N^{\cdot 5} = X$) for X on the X-axis as in Fig. 6.3. The transformation to the square root scale is particularly useful with fertilizer-yield functions because linear regressions, on the square root scale with the form

$$Y = a_0 + a_1N^{\cdot 5}$$

give a better representation of the typically curved fertilizer-yield relationship than the corresponding regressions on the natural scale with the form

$$Y = a_0 + a_1N$$

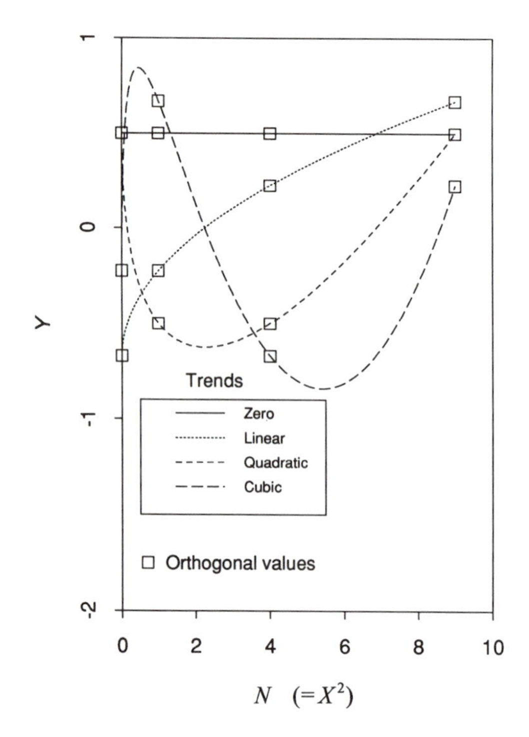

Fig. 6.3 Orthogonal polynomial trends for N = 0, 1, 4, 9 *on the square root scale* derived from Fig. 6.1 by substituting $N = X^2$.

Since the regression sum of squares in analyses of variance for these linear regressions is p_1^2 as noted above, this means that the linear trend coefficient p_1 in the linear trend term p_1L of orthogonal polynomial regressions on the square root scale is larger and contains more information about yield response to the fertilizer treatments than the corresponding coefficient for orthogonal polynomial regressions on the natural scale. This makes p_1 values for linear trends on the square root scale that have been estimated from the data of fertilizer experiments more useful for studies on relationships between the results of the experiments and site variables than corresponding values for linear trends on the natural scale. This fact is utilized in the development of general soil fertility models relating the results of experiments to site variables, to be described in chapters 8 and 9.

#6.6 Multifactor polynomials

The orthogonal polynomial procedures for obtaining regressions and analyses of variance for single factor polynomials can be extended for multifactor polynomials using a matrix obtained by the Kronecker or direct product of matrices for the different factors. Thus for example if data have been obtained from an experiment with a factorial design for the effects of treatments for the three factors R, S and T on Y, the polynomial regression

$$Y = b_0 + b_1R + b_2S + b_3T + b_4RS + b_5RT + b_6ST + b_7R^2 + b_8S^2 + b_9T^2 \quad (6.16)$$

can be estimated from the data by the regression procedure $\mathbf{b} = (\mathbf{X}'\mathbf{X})^{-1}\mathbf{X}'\mathbf{y}$ and the same regression in orthogonal form

$$Y = p_0Z + p_1L_r + p_2L_s + p_3L_t + p_4L_rL_s + p_5L_rL_t + p_6L_sL_t + p_7Q_r + p_8Q_s + p_9Q_t \quad (6.17)$$

can be obtained by $\mathbf{p} = \mathbf{E}'\mathbf{y}$ where $\mathbf{E}$ has now been obtained by selecting appropriate columns from the matrix $\mathbf{D}$ formed by the direct product of matrices[2] of orthogonal values for R, S and T

$$\mathbf{D} = \mathbf{R} * \mathbf{S} * \mathbf{T} \quad (6.18)$$

Equivalently if data have been obtained from a fertilizer experiment for the effects of the three nutrient rates *N*, *P* and *K* on crop yield *Y*, the polynomial regression on the square root scale

$$Y = b_0 + b_1N^{\cdot 5} + b_2P^{\cdot 5} + b_3K^{\cdot 5} + b_4(NP)^{\cdot 5} + b_5(NK)^{\cdot 5} + b_6(PK)^{\cdot 5} + b_7N + b_8P + b_9K$$

can be estimated by the $\mathbf{b} = (\mathbf{X}'\mathbf{X})^{-1}\mathbf{X}'\mathbf{y}$ procedure, and the same regression in orthogonal form

$$Y = p_0Z + p_1L_n + p_2L_p + p_3L_k + p_4L_nL_p + p_5L_nL_k + p_6L_pL_k + p_7Q_n + p_8Q_p + p_9Q_k$$

obtained by $\mathbf{p} = \mathbf{E}'\mathbf{y}$, by substituting $R = N^{\cdot 5}$, $S = P^{\cdot 5}$, $T = K^{\cdot 5}$ in the above formulas. Analyses of variance for such regressions are obtained as before (Table 6.5) by squaring the coefficients of the orthogonal

[2] Searle (1966) for a full description and Colwell (1978) for a Fortran computer program.

trends or by using analyses of variance for an appropriate series of non-orthogonal regressions as will be described in chapter 7.

An example of an analysis of variance with data from a factorial fertilizer experiment is illustrated in Fig. 6.4 with the square root quadratic[3]

$$Y = 147.7 + 12.34N^{.5} + 19.29P^{.5} + 13.28(NP)^{.5} + 0.213N - 15.12P \tag{6.19}$$

and the corresponding orthogonal regression

$$Y = 6801Z + 1912L_n + 3111L_p + 997.9L_nL_p + 19.3Q_n - 651.7Q_p \tag{6.20}$$

estimated from the data of a fertilizer experiment, Y being the yield of wheat and the N and P being rates of application of nitrogen and phosphorus, all as kg/ha. The analysis of variance, obtained by squaring the coefficients of the orthogonal regression (6.20) separates components of the regression sum of squares illustrated by the 3-dimensional graphs labelled p_0Z, p_1L_n, etc. The sum of these components, indicated in the figure by

$$p_0Z + p_1L_n + p_2L_p + p_3L_nL_p + p_4Q_n + p_5Q_p = Y$$

gives the surface for the regression (6.19). The surfaces for the zero, linear and quadratic trends correspond to the trends on the square root scale, illustrated in Fig. 6.3 for single factor factorials, but the linear × linear interaction trend $p_3L_nL_p$ is a new feature required for the analysis of multinutrient fertilizer-yield functions.

#6.7 Relationships between orthogonal and non-orthogonal polynomial regressions

Equations relating the coefficients p_0, p_1, ... for orthogonal regressions and b_0, b_1, ... for non-orthogonal polynomial regressions are derived from the formulas for the polynomial expressions that serve as regressor variables in the orthogonal polynomial regressions. Thus conversion equations are derived here for single factor polynomials of

---

[3] Based on an example in Colwell (1977) derived from a factorial experiment with wheat and treatments for the fertilizer nutrients N and P.

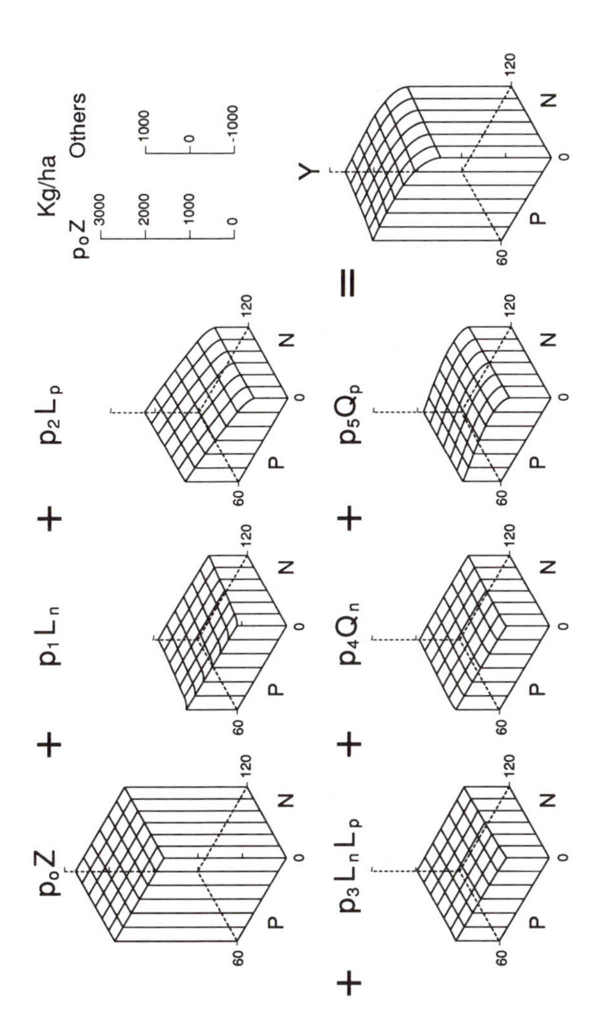

Fig. 6.4 Graphical representation of the analysis of variance of the regression (6.19) into orthogonal components (reproduced from Colwell, 1977).

degree 0, 1 and 2 using the formulas (6.3) and corresponding equations can be derived similarly for higher degree and multivariate polynomials as detailed elsewhere (Colwell 1978).

Polynomials, degree 0

The conversion equation for the zero degree polynomial $L = p_0 Z$ follows simply from the definition $Z = 1/n^{.5}$

$$\overline{Y} = g_{00} p_0 \quad \text{where} \quad g_{00} = 1/n^{.5} \tag{6.21}$$

Polynomials, degree 1

Coefficients for regressions in the alternative forms

$$Y = a_0 + a_1 X$$

and

$$Y = p_0 Z + p_1 L$$

are related by

$$a_0 = g_{00} p_0 + g_{01} p_1 \quad \text{and} \quad a_1 = g_{11} p_1 \quad \text{where}$$

$$g_{00} = \frac{1}{n^{.5}},\ g_{01} = \frac{-\overline{X}}{(\Sigma x^2)^{.5}},\ g_{11} = \frac{1}{(\Sigma x^2)^{.5}} \text{ and } \Sigma x^2 = \sum_{i=1}^{n} (X_i - \overline{X})^2 \tag{6.22}$$

Polynomials, degree 2

Coefficients for regressions in the alternative forms

$$Y = b_0 + b_1 X + b_2 X^2$$

and

$$Y = p_0 Z + p_1 L + p_2 Q$$

are related by

$$b_0 = g_{00} p_0 + g_{01} p_1 + g_{02} p_2$$

$$b_1 = g_{11} p_1 + g_{12} p_2$$

$$b_2 = g_{22} p_2$$

where g_{00}, g_{01} and g_{11} have the values defined above and

$$g_{02} = a_2^* - b_2^* \overline{X} + c_2^* \overline{X}^2, \quad g_{12} = b_2^* - 2c_2^* \overline{X} \quad \text{and} \quad g_{22} = c_2^* \tag{6.23}$$

with the values for a_2^*, b_2^* and c_2^* as defined for a_2, b_2 and c_2 in (6.6), the asterisks being added to avoid notational confusion with the present regression coefficients.

Polynomials, degree 3

The same general procedure is followed with the cubic regressions

$$Y = c_0 + c_1X + c_2X^2 + c_3X^3$$

and

$$Y = p_0 Z + p_1 L + p_2 Q + p_3 C$$

to obtain conversion equations of the form

$$c_0 = g_{00}p_0 + g_{01}p_1 + g_{02}p_2 + g_{03}p_3$$

$$c_1 = g_{11}p_1 + g_{12}p_2 + g_{13}p_3$$

$$c_2 = g_{22}p_2 + g_{23}p_3$$

$$c_3 = g_{33}p_3 \tag{6.24}$$

Multiple variable quadratics

Conversion equations are derived similarly for multifactor polynomials. For example the coefficients of the square root quadratics

$$Y = b_0 + b_1N^{.5} + b_2P^{.5} + b_3(NP)^{.5} + b_4N + b_5P$$

and

$$Y = p_0Z + p_1L_n + p_2L_p + p_3L_nL_p + p_4Q_n + p_5Q_p$$

as used for two nutrient fertilizer-yield functions, are related by equations of the form

$$b_0 = g_{00}p_0 + g_{01}p_1 + g_{02}p_2 + g_{03}p_3 + g_{04}p_4 + g_{05}p_5$$

$$b_1 = g_{11}p_1 + g_{13}p_3 + g_{14}p_4$$

$$b_2 = g_{22}p_2 + g_{23}p_3 + g_{15}p_5$$

$$b_3 = g_{33}p_3$$

$$b_4 = g_{44}p_4$$

$$b_5 = g_{55}p_5 \tag{6.25}$$

with values for the gs now determined by the direct product of matrices for the N and P as well as equations (6.4) to (6.6). However for the applications of these relationships to be described in following chapters, actual values for the gs do not have to be determined. Rather the relationships are used simply to indicate direct relationships in which the orthogonal trend coefficients and appropriately selected coefficient of non-orthogonal regressions differ only by constants, as for example in (6.25), $b_3 = g_{33}p_3$, $b_4 = g_{44}p_4$ and $b_5 = g_{55}p_5$.

6.8 Applications of the relationships

Yield variables

The above relationships between coefficients for orthogonal and non-orthogonal polynomials can be used to define *yield variables* that can be estimated directly by ordinary polynomial regressions and used to establish relationships between the results of fertilizer experiments and site variables in variable regions. Thus if the series of polynomial regressions

$$Y = \underline{\bar{Y}}$$

$$Y = a_0 + \underline{a_1} N^{.5} + \underline{a_2} P^{.5}$$

$$Y = b_0 + b_1 N^{.5} + b_2 P^{.5} + \underline{b_3} (NP)^{.5} + \underline{b_4} N + \underline{b_5} P \qquad (6.26)$$

are estimated from the data of a suitably designed fertilizer experiment, then the relationships (6.21) to (6.25) show that the mean and the underlined coefficients in these regressions relate directly to the coefficients of the orthogonal polynomial trends in the orthogonal polynomial

$$Y = p_0 Z + p_1 L_n + p_2 L_p + p_3 L_n L_p + p_4 Q_n + p_5 Q_p$$

with the conversion equations $\bar{Y} = g_{00}p_0$, $a_1 = g_{11}p_1$, $a_2 = g_{22}p_2$, $b_3 = g_{33}p_3$, $b_4 = g_{44}p_4$ and $b_5 = g_{55}p_5$. Since the g_{00}, g_{11}, ... g_{55} are constants, estimates of the mean and these particular coefficients for a series of experiments can be used to represent the orthogonal polynomial components of analyses of variance of regressions of the form

$$Y = b_0 + b_1 N^{.5} + b_2 P^{.5} + b_3 (NP)^{.5} + b_4 N + b_5 P$$

and hence to establish relationships with site variables. These types of relationships serve to indicate regression coefficients that can be used to derive *general soil fertility models* as described in chapters 8 and 9.

Relationships between coefficients for polynomials of different degree

The conversion equations also indicate ways for calculating the regression coefficients for polynomials of lower degree from polynomials of higher degree. Thus given the polynomial regression

$$Y = b_0 + b_1X + b_2X^2$$

the coefficient a_1 in the lower degree regression

$$Y = a_0 + a_1X$$

is calculated by

$$a_1 = b_1 + b_2h$$

where

$$h = \frac{\Sigma x^3}{\Sigma x^2} + 2\overline{X}$$

$$\text{and} \quad \Sigma x^2 = \sum_{1=1}^{n}(X_i - \overline{X})^2 \quad \text{and} \quad \Sigma x^3 = \sum_{1=1}^{n}(X_i - \overline{X})^3 \qquad (6.27)$$

Similarly given the regression coefficients for the square root quadratic

$$Y = b_0 + b_1N^{.5} + b_2P^{.5} + b_3(NP)^{.5} + b_4N + b_5P$$

the coefficients a_1 and a_2 in the lower degree polynomial

$$Y = a_0 + a_1N^{.5} + a_2P^{.5}$$

are calculated by

$$a_1 = b_1 + b_3\overline{P} + b_4h_n$$

$$a_2 = b_2 + b_3\overline{N} + b_5h_p$$

where

$$h_n = \frac{\Sigma n^3}{\Sigma n^2} + 2\overline{N}$$

$$h_p = \frac{\Sigma p^3}{\Sigma p^2} + 2\overline{P}$$

$$\Sigma n^2 = \sum_{i=1}^{n}(N_i - \overline{N})^2 \;,\; \Sigma n^3 = \sum_{i=1}^{n}(N_i - \overline{N})^3$$

$$\Sigma p^2 = \sum_{i=1}^{n}(P_i - \overline{P})^2 \text{ and } \Sigma p^3 = \sum_{i=1}^{n}(P_i - \overline{P})^3 \qquad (6.28)$$

Chapter 7
Designs for Fertilizer Experiments

7.1 General features of the designs

Experiments on the effects of fertilizers on crop production are expensive and time consuming, especially when series of experiments have to be carried out to obtain results that represent the range of soil fertility and growing conditions that occur in a variable region. For this reason designs are needed for experiments that are about as small and simple as possible, consistent with their providing the required information with an acceptable order of accuracy. Designs for fertilizer experiments have been selected for this chapter on this basis, to range in size from 16 to 64 plots, 16 being judged about the minimum for the required level of accuracy with routine experiments for the diagnosis of soil nutrient deficiencies or the estimation of optimal rates for one or two nutrients, and 64 about the maximum practicable for more elaborate experiments, as for the determination of the simultaneous requirements for four nutrients. In general the smaller designs are recommended, there being no advantage in carrying out unnecessarily large and complicated experiments - attempts to impress with large and complex designs often lead to disappointment. The recommended designs, ranging in size from 16 to 64 plots, are thus judged to be practicable and safe, given reasonable care and the conditions commonly encountered with field experiments.

All of the recommended designs are based on standard factorial or fractional factorial designs as listed in standard references[1]. Factorial designs consist of all combinations of two or more treatments for each of a series of factors, the factors for fertilizer experiments being

[1] In particular from those given by Cochran and Cox (1957), John and Quenouille (1977) and Mead (1988).

fertilizers or fertilizer nutrients and the treatments being rates of application of these fertilizers or nutrients. Such designs are preferred for fertilizer experiments for several reasons:

1. Treatment rates can be chosen to produce data that provide efficient regression estimates of polynomial functions for the fertilizer-yield relationship.
2. The factorial combination of treatment rates provides data suitable for the estimation of interaction effects such that the effect of applications of each nutrient may be affected by applications of others.
3. Complete factorial designs or certain fractional factorial designs produce data that are suitable for direct and detailed analyses of variance into orthogonal polynomial components. These are needed for studies on the relationships between the results of series of experiments and site variables.
4. Fractional designs can be used for larger experiments to increase the efficiency and accuracy of estimates of important treatment effects.
5. Treatments can be allocated to small blocks to increase the accuracy of the results by separating the effects of site variability from treatment effects in analyses of variance.

7.2 Choice of treatments

Factorial designs are conveniently specified by coding the treatments with sets of integers 0, 1, 2, ... for each factor as described in chapter 2 (Table 2.1) and appropriate nutrient treatment rates have to be chosen to correspond with the integers. Firstly highest and lowest treatment rates have to be chosen to define the range to be represented by the experimental data and then intermediate rates have to be chosen with an appropriate relative spacing, when there are more than these two rates. It is important to note in this regard that the evenly spaced integer code values 0, 1, 2, ... do not necessarily imply an even spacing of treatment levels and in particular that the integer value 0 should not be taken to necessarily imply a zero treatment rate. For example the codes 0, 1, 2, 3 could correspond to the set of treatment rates 0, 10, 20, 40 kg P/ha or alternatively to the set 4, 22, 54, 100 kg N/ha.

Range for treatment rates

The traditional range of treatment rates chosen for fertilizer experiments is from a lowest rate of zero to a highest rate that is judged sufficient to eliminate any likely soil deficiency, with the objective of obtaining data showing maximum responses to the respective nutrient applications. This is appropriate for diagnostic or exploratory experiments that have the purpose of detecting or demonstrating soil nutrient deficiencies but is less appropriate if the purpose of the experiments is to estimate optimal fertilizer rates. For example although an application of 500 kg N/ha might be judged sufficient to eliminate any N nutrient deficiency for a particular crop and soil, so that treatments 0 and 500 kg N/ha would be appropriate to demonstrate the magnitude of effects due to soil N deficiency, the range would be grossly excessive if the purpose was to estimate optimal application rates when previous experience had indicated that optimal rates are likely to be somewhere within the range 20 to 100 kg N/ha. Under such circumstances the use of the treatment range 0 to 500 kg N/ha would be likely to distort estimates of the fertilizer-yield relationship from experiment data in the important range 20 to 100 kg N/ha, producing interpolation errors and hence misleading estimates of the optimal rate. Rather, the narrower range 10 to 150 kg N/ha would be better, that is a treatment range that is a little wider than the important range of 20 to 100 kg N/ha. As noted previously, the yield produced by the zero application rate is not required for the calculation of optimal rates and the coding 0 for the lowest treatment rate in experimental designs does not necessarily correspond to the zero rate. A "rule-of-thumb" in this regard is to choose a lowest rate that is about ½ that judged to be sufficient to replace the nutrient removed by an average crop, that is about half the rate required to maintain the existing level of soil nutrient with normal cropping. Maintenance rates for this purpose should be determined by analyses of the crops produced in the region of the experiment since crop compositions vary widely for different regions and growing conditions. A lowest rate of ½ the maintenance rate estimated in this way is expected to allow for a possible overestimation such as may result from an excessive use of fertilizers and a consequent excessive nutrient consumption by crops, as in highly developed agricultural regions.

Relative spacing for intermediate treatment rates

When experiments have the purpose of providing data for the estimation of polynomial functions, treatment rates can be chosen to coincide as closely as possible with the contrasts or extremes of their orthogonal components, as shown by the equations for orthogonal polynomials (6.3). Such extremes are illustrated in Fig. 7.1 by graphs for zero, linear and quadratic orthogonal components of polynomial regressions with the form

$$Y = b_0 + b_1 X + b_2 X^2 \tag{7.1}$$

as defined by the equations (6.3) for the zero, linear and quadratic orthogonal trends Z, L and Q in the corresponding orthogonal regression

$$Y = p_0 Z + p_1 L + p_2 Q \tag{7.2}$$

The positions of the polynomial values giving the orthogonal contrasts for three evenly spaced treatments is indicated by asterisks on the graphs. Similar graphs and asterisks for four evenly spaced treatments are given in Fig. 7.2 for the zero, linear, quadratic and cubic trends of the cubic polynomial

$$Y = b_0 + b_1 X + b_2 X^2 + b_3 X^3 \tag{7.3}$$

or the corresponding orthogonal polynomial

$$Y = p_0 Z + p_1 L + p_2 Q + p_3 C \tag{7.4}$$

These *same* polynomial trends and asterisks for orthogonal values for three and four evenly spaced treatments on the natural scale are illustrated in Figs 7.3 and 7.4 for polynomials on the square root scale[2], that is for

$$Y = b_0 + b_1 N^{\cdot 5} + b_2 N \tag{7.5}$$

and

$$Y = b_0 + b_1 N^{\cdot 5} + b_2 N + b_3 N^{1 \cdot 5} \tag{7.6}$$

The graphs in these figures show two important features for the choice of treatment rates:

1. They show the relative magnitudes of the orthogonal polynomial values for the treatment rates in relation to the polynomial regression components that they estimate and hence the relative importance of the various treatments for the estimation of each of the components of polynomial regressions. Since the magnitudes of

---

[2] Trend values for these graphs are obtained by the substitution of $N = X^2$ (because $N^{\cdot 5} = X$) giving the range 0 to 100 for N and the relative spacings 0:1:4 and 0:1:4:9 in place of 0 to 10, and 0:1:2 and 0:1:2:3 for X.

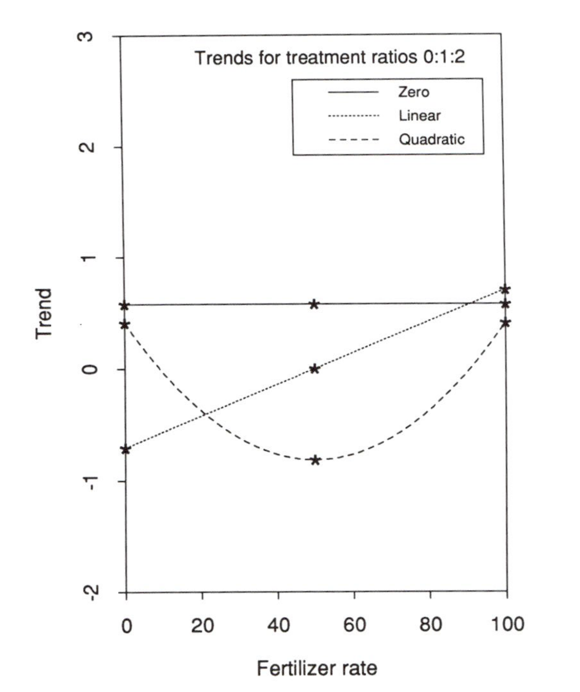

Fig. 7.1 Trends for orthogonal polynomial equations (7.3) with * for orthogonal polynomial values, for three treatment rates in ratio 0:1:2.

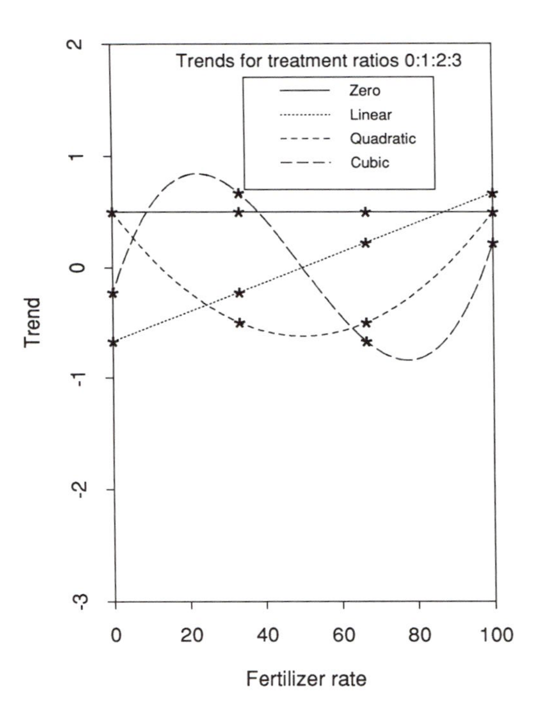

Fig. 7.2 As for Fig. 7.2, for four treatment rates in ratio 0:1:2:3.

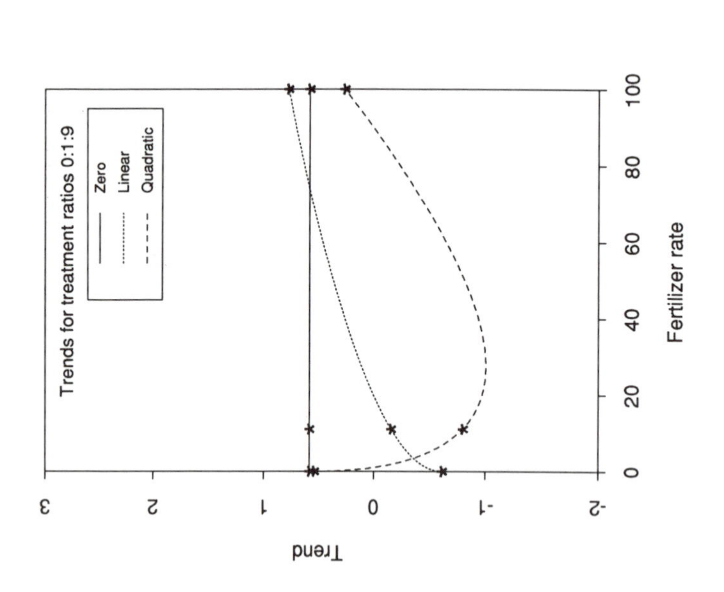

Fig. 7.3 Trends for orthogonal polynomial equations (7.3) with * for orthogonal polynomial values on the square root scale, for three treatment rates in ratio 0:1:4 (0:1:2 on the square root scale).

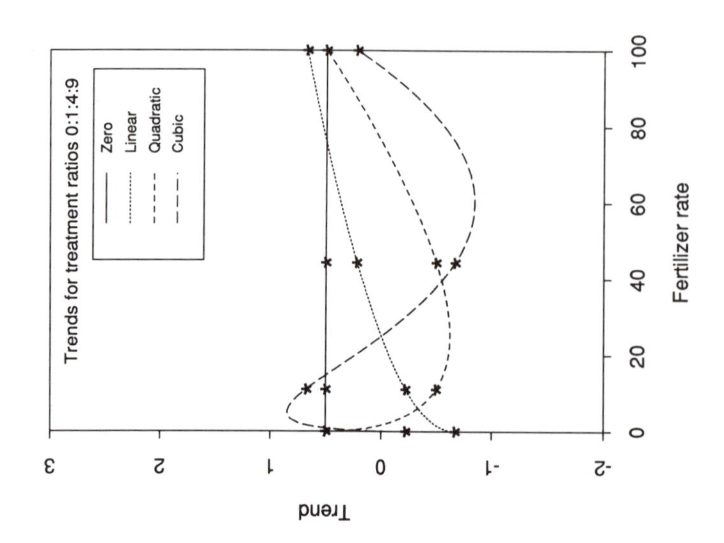

Fig. 7.4 As for Fig. 7.3, for four treatment rates in ratio 0:1:4:9 (0:1:2:3 on the square root scale).

the coefficients for the orthogonal regressions are determined by vector products, $\mathbf{p}_i = \mathbf{x}_i'\mathbf{y}$ where $\mathbf{x}_i$ is the vector of orthogonal polynomial values and $\mathbf{y}$ is the yield vector (equation 6.12), the greater the polynomial value for a treatment rate the greater the contribution of the yield data produced by the treatment to the estimate of the trend coefficient $\mathbf{p}_i$, that is to the estimate of the orthogonal component of the polynomial. For example the intermediate treatment rate in Figs 7.1 and 7.3 gives no information about the linear trend because the trend has the value zero for this rate. Similarly the intermediate treatment rates in Figs 7.2 and 7.4 are not well placed to measure the cubic trend.

2. They show the nature of the interpolation between the treatment rates for each component trend. Where interpolated values are large, that is when the treatment rates are not close to the peaks of the trends, regression estimates will be particularly vulnerable to interpolation error. For example, interpolation error effects associated with the cubic trend are likely to be large with the rates in Figs 7.2 and 7.4 because the treatment rates are not close to the peaks for this trend. Consequently graphs of cubic regressions estimated from regressions with the relative treatment spacings 0:1:2:3 and 0:1:4:9 are likely to show unrealistic curved effects between the treatment rates for the codes 0 and 1 and for 3 and 4.

The features of the graphs for orthogonal polynomial trends in Figs 7.1 and 7.2 thus provide bases for the evaluation of treatment rates with the relative spacings 0:1:2 and 0:1:2:3 for the estimation of polynomial regressions on the natural scale and similarly Figs 7.3 and 7.4 provide bases for the evaluation of treatment rates with the spacings 0:1:4 and 0:1:4:9 for the estimation of polynomials on the square root scale.

Zero trends

The graphs for the zero trend component are straight lines parallel to the X-axis indicating that spacings have no effect on the contributions of yield data to estimates of the zero trend coefficient. Since the zero trend "regression" $Y = p_0 Z$ is simply the equation of the mean $Y = \overline{Y}$, this simply corresponds to the fact that all yield data are equally important for the estimation of the mean.

Linear trends

The linear trend graphs show that the polynomial values are largest at the extremes of the range and consequently that extreme treatment rates produce data that have the greatest effect on the estimation of the linear trend $Y = p_1L$ and hence of this component of polynomial regressions. The extreme treatment rates are particularly important in this respect when there are three treatment levels with the respective relative spacings 0:1:2 on the natural scale (Fig. 7.1) and 0:1:4 on the square root scale (Fig. 7.3) since, as already noted, in these cases the orthogonal polynomial value for the centre rate is zero. Thus if a relationship is linear for the range of treatment rates, the regressions

$$Y = b_0 + b_1 X$$

$$Y = b_0 + b_1 N^{.5}$$

are best estimated with only two treatment rates, coinciding with the extremes of the range for the relationship.

Quadratic trends

The quadratic components have a peak midway between the treatment extremes (midway on the square root scale for Figs 7.3 and 7.4) and the quadratic effect $Y = p_2Q$ is consequently best estimated by treatment rates located at the extremes and mid point of the range. Thus if a relationship is quadratic for the range of the treatment rates, the regressions

$$Y = b_0 + b_1X + b_2X^2$$

$$Y = b_0 + b_1N^{.5} + b_2N$$

are best estimated with three treatment rates with the relative spacings 0:1:2 for polynomials on the natural scale and 0:1:4 for polynomials on the square root scale. Note however that with four treatment rates the relative spacings 0:1:2:3 for the natural scale and 0:1:4:9 for the square root scale give orthogonal polynomial values close to this ideal, the values for the centre two treatment rates being close to the peak. Although the peak is estimated in these cases by interpolation, the magnitude of misleading interpolation effects (interpolation error) is likely to be small and negligible.

Cubic trends

Cubic trends have two peaks as illustrated in Figs 7.2 and 7.4 so that ideal treatments for the estimation of this type of effect are those that give orthogonal polynomial values located at the peaks as well as at the extremes of the treatment range. The spacings 0:1:2:3 and 0:1:4:9 for four treatment levels produce orthogonal polynomial values that only approximate to this ideal so that the peaks are estimated with some interpolation. If a cubic effect is to be estimated, treatments will be located ideally at the extremes but given the requirement of a centre rate for the estimation of the quadratic effect, this would require a total of five treatment levels.

Spacings 0:1:2:3 and 0:1:2:4

The suitability of other treatment spacings for the estimation of polynomial components of polynomials on the *square root scale* may be examined similarly by calculating orthogonal polynomial equations as illustrated in Figs 7.5 and 7.6 for the popular relative treatment spacings 0:1:2:3 and 0:1:2:4.

The spacing 0:1:2:3 approximates reasonably closely to the ideal for the estimation of quadratic or cubic regressions on the natural scale as noted above. But Fig. 7.5 shows that this spacing is not satisfactory for the estimation of polynomial regressions on the square root scale since the first portion of the cubic component, between the lowest pair of treatment rates, estimates the first peak with a very large interpolation. Consequently estimates from cubic polynomial regressions between these treatment levels will be particularly vulnerable to an interpolation error due to their cubic component. The quadratic peak is also estimated by a substantial interpolation between the lowest pair of treatment rates. The relative treatment spacing 0:1:2:3 should consequently not be used for the estimation of polynomials on the square root scale.

The graphs for the 0:1:2:4 relative treatment spacings in Fig. 7.6 show similar though smaller interpolation hazards. The interpolation is still large for the first cubic peak but not serious for the quadratic peak because the polynomial value for the second treatment rate is close to the peak and the value for the third rate is also substantial. Thus this spacing is reasonably satisfactory for the estimation of quadratics on the square root scale, particularly if the estimated quadratic effects are small as is often the case with fertilizer experiments[3]. However it is not satisfactory for the estimation of cubic trends on this scale nor consequently for the estimation of cubic polynomials.

Ideal treatment rates

The above considerations concerning the range of treatment rates and their relative spacings indicate bases for calculating treatment rates

---

[3] See for example the analyses of variance for a large number of fertilizer experiments described by Colwell (1977).

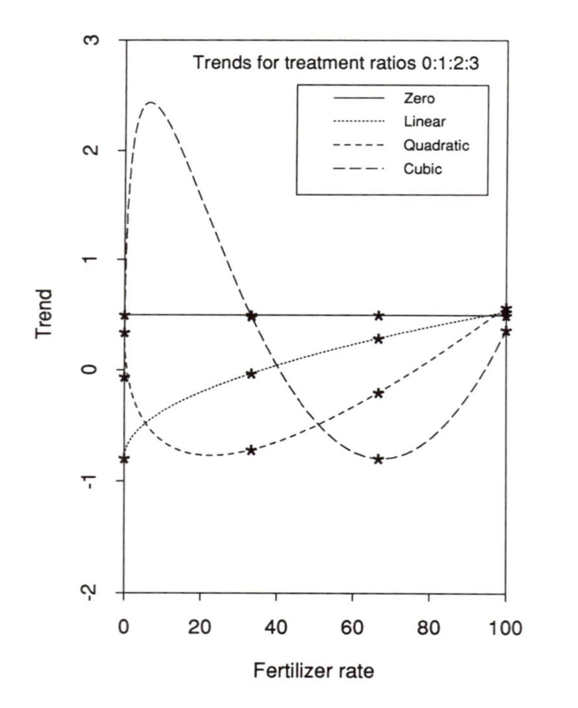

Fig. 7.5 Trends for orthogonal polynomial equations (7.3) with * for orthogonal polynomial values, on the square root scale, for three treatment rates in ratio 0:1:2:3.

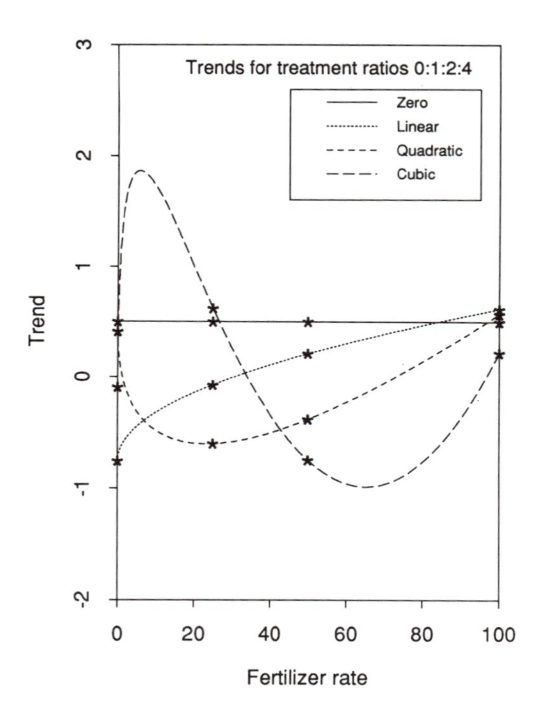

Fig. 7.6 As for Fig. 7.5, for four treatment rates in ratio 0:1:2:4.

that are ideal for the estimation of the component trends of polynomial regressions. Thus for polynomials on the square root scale, if treatment rates are to range from a lowest value L to a highest value H, the three ideal treatment rates N_0, N_1, N_2 corresponding to the codes 0, 1, 2 in designs, are calculated by

$$N_0 = L \ , \quad N_1 = \frac{\left(L^{.5} + H^{.5}\right)^2}{4} \quad \text{and} \quad N_3 = H \qquad (7.7)$$

Similarly if four treatment rates are required to estimate a square root quadratic and also to check on the assumed unimportance of the cubic effect by analysis of variance, the four evenly spaced rates

$$N_0 = L \ , \quad N_1 = \frac{\left(2L^{.5} + H^{.5}\right)}{9} \ , \quad N_2 = \frac{\left(L^{.5} + 2H^{.5}\right)}{9} \quad \text{and} \quad N_3 = H \quad (7.8)$$

are ideal. Example sets of ideal rates calculated with these formulas are given in Table 7.1 using in this instance the coding N_0, N_1, N_2 and N_3 rather than the simpler 0, 1, 2 and 3. When a set of constant treatment rates have to be chosen for a series of experiments in a region where soil nutrient levels range from sufficiently high not to warrant the application of any nutrient to low, warranting a high rate H, treatments proportional to the sets 1 or 4 in this table would be ideal. An alternative to set 4 often preferred to avoid the excessively low "ideal" rate N_1 has the relative spacing 0:1:2:4 giving for example the treatment rates 0, 25, 50, 100 rather than 0, 11.11, 44.44, 100. As noted with Fig. 7.6 the 0:1:2:4 relative spacing is reasonably satisfactory for the estimation of square root quadratic functions but not for the estimation of cubic trends or square root cubic functions.

Table 7.1 Example ideal fertilizer treatment rates for the estimation of square root quadratic fertilizer-yield functions.

Treatment set	N_0	N_1	N_2	N_3
1	0	25	100	
2	1	30.25	100	
3	4	36	100	
4	0	11.11	44.44	100
5	1	16	49	100
6	4	21.78	53.78	100

7.3 Blocks

Allocation of treatments to blocks

The residual mean square in analyses of variance can be reduced, and hence the accuracy of estimates of treatment effects increased, by the allocation of treatments to blocks in the layout of experiments as explained in chapter 2, especially if the blocks can be chosen to coincide with potential sources of within site variation, for example with adjacent terraces on steep slopes. The selection of treatments for blocks in designs, called blocking, is done on a mathematical basis so that block effects can be separated from treatment effects in an analysis of variance, thus reducing loss of information about the treatment effects. With small experiments consisting of replicates of relatively few treatments, this is accomplished simply by replicating the treatments in blocks. However for experiments with many treatments, the simple replication of all treatments in blocks produces large blocks which, because of their size, may do little towards reducing effects from site variability. For such experiments small blocks are produced by allocating appropriate selections of the treatments to blocks in such a way that treatment effects will be independent of block effects in analyses of variance and that contrasts for unimportant or meaningless treatment effects are confounded with the block effects, as described in standard texts on experimental design. The blocked designs described below have been selected in this way, so that treatment effects corresponding to the orthogonal components of polynomial fertilizer-yield functions are independent of the block effects and unlikely effects are confounded with the block effects.

Dummy variables for blocks

Block effects due to differences in growing conditions within an experiment site may be regarded as being equivalent to treatment effects, the "treatment" being "applied" by the location of experiment plots in the blocks. Moreover the "treatments" can be given numerical values and represented by variables in regressions by defining convenient dummy variables such as the L_b, Q_b and C_b in Table 7.2, corresponding to orthogonal polynomial values for an evenly spaced variable. For example with fertilizer experiments with treatments for application rates N and four blocks, the experimental results may be represented by regression equations of the form

$$Y = b_0 + b_1 N^{.5} + b_2 N + b_3 L_b + b_4 Q_b + b_5 C_b \qquad (7.9)$$

Table 7.2 Dummy variables for linear, quadratic and cubic trends amongst blocks in experiments.

	2 blocks	3 blocks		4 blocks		
Block	L_b	L_b	Q_b	L_b	Q_b	C_b
1	−1	−1	+1	−3	+1	−1
2	+1	0	−2	−1	−1	+3
3		+1	+1	+1	−1	−3
4				+3	+1	+1

where L_b, Q_b and C_b are regressor variables respectively for linear, quadratic and cubic trends in yield levels across the four blocks.

The use of orthogonal polynomial values for block variable trends has two useful consequences:

1. The regression equation for treatment effects over an experiment site considered as a whole is derived simply by deleting the block terms from regressions since the mean value for each variable is zero. Thus with the regression (7.9) treatment effects averaged over blocks are represented by

$$Y = b_0 + b_1 N^{.5} + b_2 N$$

2. The values for the dummy variable regression coefficients provide independent measures of the magnitude of the respective trends across the blocks.

Nearest neighbour methods

Nearest neighbour methods[4] have been developed to allow and adjust data for variations within experiment sites. In general terms the procedures are based on the estimation of site effects by the subtraction of mean yields for each treatment from the observed data yield and then the adjustment of the yield of each plot by the residuals for the neighbouring plots. The basic assumption in this procedure is that the factors producing "error" effects on individual plots have similar effects on the neighbouring plots, independent of the effects of the treatments for each plot. The assumption seems reasonable for many types of experiments such as plant breeding experiments in which

[4] Attributed originally to Papadakis (1937) and Bartlett (1938) - see review by Kempton and Howes (1981). A more sophisticated procedure is described by Williams (1986).

yields are compared for different breeding lines of a crop species. The assumption is very doubtful however with fertilizer experiments since the effects of some site factors, particularly of soil nutrient level or of factors that are affected by soil nutrient level, may differ with the fertilizer treatment for each plot. For example if there are within site variations in soil phosphorus and if the plots receive phosphorus fertilizer treatments, then effects of the soil phosphorus may be expected to be much greater for plots receiving the zero treatment rate than for those receiving the highest phosphorus treatment rate. Similarly it is a common observation that some disease and weed effects are reduced by fertilizer applications so that effects of these factors will be greater for plots receiving low fertilizer treatment rates.

Such considerations indicate the need to allow for the basic features of fertilizer experiments before adopting procedures developed for other types of experiment. In any case since designs requiring only small blocks can be used for fertilizer experiments, the need for nearest neighbour procedures is very much less than with experiments requiring large blocks for which the procedures were originally developed.

7.4 Fractional designs

Complete factorial designs, consisting of all factorial combinations of treatment rates for each nutrient (factor) are *orthogonal* in the sense that with n treatment combinations, n treatment contrasts can be defined that are orthogonal with respect to each other. With fractional factorial designs consisting of only an appropriately selected fraction of the factorial combinations, some of the treatment contrasts are *confounded* with each other so that some treatment effects cannot be distinguished from others. This confounding is useful in experiment designs if one or both of the confounded effects is only likely to represent error aberrations amongst the data rather than meaningful treatment effects. Thus confounding of unimportant or meaningless effects is used to reduce the size of experiments, making them more efficient for measuring meaningful effects. In particular, fractional designs that have been developed in this way can be expected to give more accurate experiments because small experiments are more easily managed and more easily located on uniform sites and within uniform blocks.

Orthogonality between treatment contrasts is shown when the correlation coefficient between the contrasts is $r = 0$ and a complete confounding of contrasts is shown when the correlation coefficient value is $r = 1$. In some designs there is a *partial confounding* between these extremes of nil and complete confounding as shown by intermediate correlations for contrasts, $0 < r < 1$, with a corresponding intermediate loss of information about the respective treatment effects. These are noted with the designs given below since any degree of confounding causes a loss of information about the effects measured by the contrasts, ranging from small and negligible when $r \to 0$ to large and serious when $r \to 1$. The correlations noted are all for evenly spaced treatment rates, corresponding to the ideal rates given in Table 7.1. With fractional designs, the non-even spacing of treatments may produce other correlations, including some non-zero correlations for some of the important treatment effects so that regression procedures must be used for analyses of variance. For the selected designs losses of information due to such confounding is unimportant since it involves either unlikely treatment effects or only small correlations for the important effects.

The choice of fractional or blocked designs on these bases depends essentially on the identification of the components of polynomial functions as described in chapter 6 and then the choice of treatments that give orthogonal contrasts for the measurement of these contrasts. It is important to understand in this regard that fractional designs have been derived by mathematical studies on orthogonality and confounding between contrasts in the designs and not, as sometimes mistakenly believed, by the identification and deletion of seemingly "unimportant" treatments indicated by the simple inspection of the treatments.

7.5 Randomization

The importance of randomizing the location of treatments in experiments has already been stressed (section 2.1) and is mentioned again because of the temptation for experimenters to simply copy the tidy layouts used to specify experimental designs as in the tables that follow, in order to produce correspondingly tidy field layouts for experiments. As explained in section 2.1, when a design and treatments have been chosen for an experiment or series of experiments, the location of the treatments within experimental blocks or sites should be randomized, as by allocating treatments to plots with random numbers. Also different randomizations should be used for each block within experiments when treatments are replicated in blocks, and for each experiment within a series of experiments.

7.6 Diagnostic Experiments

Diagnostic experiments have the simple purpose of detecting and identifying soil nutrient deficiencies and hence of providing bases for the planning of more comprehensive experiments to determine fertilizer requirements. Consequently simple designs may be used with two treatment rates for each nutrient, consisting of a zero rate and a high rate judged sufficient to correct any likely deficiency. Since the yield response to any treatment may be affected by levels of application of other nutrients, factorial combinations of the treatment rates are required to show interaction effects with, incidentally, the advantage of a consequent "inbuilt" replication to allow for error effects. The standard 2^n series of factorial designs, as listed by Cochran and Cox (1957), are suitable for such experiments, as with the following 2^3 design for three nutrients and the 2^4 design for four nutrients.

Analyses of variance for treatment and block effects for the 2^n series of factorials may be obtained using the desk calculator procedures described in standard text books on experimental design or alternatively by using analyses of variance and tests of significance derived from the regressions as described below with the 2^3 and 2^4 designs. Given a computer facility, this latter procedure is more convenient because it can also be used when there has been a loss of data and because it gives regression equations to represent the effects indicated by analyses of variance. These can be used for preliminary studies on fertilizer requirements, as with the example experiment in chapter 2.

Design 1 - diagnosis for three nutrients (2^3 factorial, 16 plots, four blocks)

The 2^3 factorial design in Table 7.3 is for three nutrients, denoted by N, P and K, each applied at two levels coded as 0 for a zero application rate and 1 for a sufficiency rate. The eight factorial treatment combinations are distributed in small blocks of four plots and all are replicated to give a total of 16 plots distributed amongst four blocks. The small blocks of only four plots make this design particularly suitable for difficult sites where variations within the experiment site could present a problem for experiments requiring large blocks.

The analysis of variance in Table 7.3 can be obtained using the regressions 1 and 2 with variables N, P and K for the treatment rates and the dummy variables L_b, Q_b and C_b for blocks effects since the sums of squares for each source of variation in the analysis of variance correspond to those used to test the significance of the underlined

Table 7.3 Design 1 for three nutrient diagnostic experiments (2^3, 16 plots, four blocks), regressions and analysis of variance for Design 1. Row numbers are for randomizations within the blocks.

Row	Block 1	Block 2	Block 3	Block 4
	NPK	NPK	NPK	NPK
1	000	001	000	001
2	011	010	011	010
3	101	100	101	100
4	110	111	110	111

Regression 1 $$Y = b_0 + b_1N + b_2P + b_3K + \underline{b_4}NP + \underline{b_5}NK + \underline{b_6}PK + \underline{b_7}L_b + \underline{b_8}Q_b + \underline{b_9}C_b$$

Regression 2 $$Y = a_0 + \underline{a_1}N + \underline{a_2}P + \underline{a_3}K + a_4L_b + a_5Q_b + a_6C_b$$

Source of variation		d.f.	Coefficient for sum of squares
Main effects	N	1	a_1
	P	1	a_2
	K	1	a_3
Interactions	N × P	1	a_4
	N × K	1	a_5
	P × K	1	a_6
Block effects	Linear	1	b_7
	Quadratic	1	b_8
	Cubic	1	b_9
Residue		6	(from regression 1)
Total		15	

Notable confounding: None

Table 7.4 Design 2 for four nutrient diagnostic experiments (2^4, 16 plots, two blocks) and regressions for analysis of variance. Row numbers are for randomization.

Row	Block 1	Block 2
	NPKS	NPKS
1	0001	0000
2	0010	0011
3	0100	0101
4	0111	0110
5	1000	1001
6	1011	1010
7	1101	1100
8	1110	1111

Regression 1 $Y = b_0 + b_1N^{.5} + b_2P^{.5} + b_3K^{.5} + b_4S^{.5} + \underline{b_5}(NP)^{.5} + \underline{b_6}(NK)^{.5} + \underline{b_7}(NS)^{.5} + \underline{b_8}(PK)^{.5} + \underline{b_9}(PS)^{.5} + \underline{b_{10}}(KS)^{.5} + \underline{b_{11}}L_b$

Regression 2 $Y = a_0 + \underline{a_1}N + \underline{a_2}P + \underline{a_3}K + \underline{a_4}S + a_5L_b$

Source of variation		d.f.	Coefficient for sum of squares
Main effect	N	1	a_1
	P	1	a_2
	K	1	a_3
	S	1	a_4
Interaction effect	N × P	1	b_5
	N × K	1	b_6
	N × S	1	b_7
	P × K	1	b_8
	P × S	1	b_9
	K × S	1	b_{10}
Block effect (L_b)		1	b_{11}
Residue		4	(from regression 1)
Total		15	

Notable confounding: None.

regression coefficients, $\underline{a_1}$, $\underline{a_2}$, ... $\underline{b_9}$. Thus the sum of squares for the N main effect is the reduction in sum of squares caused by the omission of the variable N in *Regression 2* corresponding to that used to test the significance of the coefficient a_1 and similarly for the other coefficients. F-ratios for tests of significance are calculated using the residual mean square of the main *Regression 1*. A corresponding procedure is demonstrated below with the example for Design 2. Alternatively and more simply, F-ratios can be calculated directly from those obtained by standard procedures for the respective coefficients of the regression, by multiplying by the factor obtained by dividing the residual mean square of each regression by that of the main regression. Since $F = t^2$ the same procedure may be used if t-tests are provided for the coefficients by a computer regression program.

The small blocks are obtained for this design by confounding the high degree interaction N × P × K with the block effects on the basis that any such interaction effect produced by the treatments is likely to be small and negligible, and that any observed effect is more likely to have been produced by an error aberration in the data. Thus the effects separated in the analysis as block effects are assumed to be due to location in the blocks with at most a negligible contribution from this higher degree interaction effect. Consequently important correlations due to the confounding in the design are indicated as *None*.

Design 2 - diagnosis for four nutrients (2^4 factorial, 16 plots, two blocks)

The selected design in Table 7.4 is similar in size to Design 1 but is for four nutrients in blocks of eight plots. Analyses of variance can be obtained with the indicated regressions following a corresponding procedure to that described for Design 1, using the sums of squares associated with the underlined coefficients. Again confounding only involves negligible interactions of higher degree.

Example: Diagnostic experiment with Design 2

An example of the use of the diagnostic Design 2 (Table 7.4) is illustrated with the experimental layout in Fig. 7.7, the treatments having been located at random in two blocks, with randomizations obtained by using random numbers 1 to 8 corresponding to the row numbers for the design in Table 7.4. Briefly, the experiment was carried out on an apparently uniform site with the purpose of identifying any deficiencies of the nutrients S, K, Mo or Zn using the treatment rates in Table 7.5 corresponding to the design coding. The yields produced by these treatments are listed for each plot in the figure.

Following the regression procedure for obtaining the analysis of variance of Table 7.4, the *Regression 1*

$$Y = 13.81 + 0.000000021S + 0.0125K + 13.33Mo + 0.850Zn + \underline{0.001563}SK + \underline{0.04167}SMo - \underline{0.0025}SZn + \underline{0.1458}KMo + \underline{0.00125}KZn + \underline{2.833}MoZn - \underline{2.688}L_b$$

and *Regression 2*

$$Y = 11.31 + \underline{0.03125}S + \underline{0.08438}K + \underline{24.58}Mo + \underline{1.225}Zn - \underline{2.688}L_b$$

were estimated from the data and used to obtain the analysis of variance in Table 7.6. Sums of squares associated with the underlined coefficients of these regressions give the values for this analysis. The sums of squares are also expressed as percentages of the total to show more clearly the relative magnitudes of the various components of the analysis. F ratios for tests of significance are calculated by dividing these sums of squares (each has one degree of freedom) by the residual mean square for the main regression (38.250 ÷ 4 = 9.5625). Alternatively F values can be obtained directly from corresponding values for tests of significance for the regression coefficients. Thus F values for the significance of the coefficients 0.001563 to –2.688 (b_5 to b_{11} of *Regression 1*) give the values for the interaction and block effects in Table 7.6 and the F values for the main effects are obtained by multiplying the F values for the coefficients 0.03125 to 1.225 (b_1 to b_4 of *Regression 2*) by the ratio of the mean squares of *Regressions 2* and *1* (7.4625 ÷ 9.5625 = 0.7804). Thus the F ratios or t^2 values for the example coefficients 1.884, 6.099, 29.152 and 20.105 multiplied by 0.7804 give the F ratios 1.47, 4.76, 22.75 and 15.69 of Table 7.6.

The analysis indicates statistically significant responses to the Mo and Zn treatments and a significant difference between the two blocks. The magnitudes of the response to each nutrient application, in the presence or absence of the other nutrients, can be calculated from *Regression 1* and the economic significance of the responses can be gauged by calculating profits as was done with the example in chapter 2 (Table 2.6). The conclusion from the example would thus be that there are soil nutrient deficiencies of Mo and Zn that warrant further more comprehensive experimentation. The reason for the significant block effect should also be investigated to help with the selection of more uniform sites for future experiments.

Table 7.5 Treatment application rates of nutrients for the example diagnostic experiment with the design of Table 7.4. L_b is a dummy variable for block effect.

Code	kg S/ha	kg K/ha	kg Mo/ha	kg Zn/ha	L_b
0	0	0	0	0	−1
1	60	40	0.3	5	+1

	Treatment Row	Treatment Code	Plot yield
Block 1	7	1101	22
	5	1000	16
	3	0100	17
	1	0001	19
	2	0010	23
	6	1011	29
	4	0111	32
	8	1110	29
Block 2	2	0011	23
	6	1010	14
	4	0110	15
	7	1100	16
	1	0000	11
	3	0101	18
	8	1111	30
	5	1001	17

Fig. 7.7 Layout of a diagnostic experiment with Design 2 (Table 7.4) and treatments of Table 7.5.

Table 7.6 Analysis of variance for the diagnostic experiment in Fig. 7.7.

Source of variation	d.f.	Sum of squares	% Sum of squares	F ratio
Main effects				
S	1	14.063	2.28	1.47
K	1	45.563	7.38	4.76
Mo	1	217.562	35.23	22.75**
Zn	1	150.063	24.30	15.69*
Interactions				
S × K	1	14.063	2.28	1.47
S × Mo	1	0.562	0.09	0.06
S × Zn	1	0.562	0.09	0.06
K × Mo	1	3.063	0.50	0.32
K × Zn	1	0.062	0.01	0.01
Mo × Zn	1	18.062	2.93	1.89
Block effect (L_b)	1	115.563	18.71	12.09*
Residue	4	38.250	6.20	
Total	15	617.438	100.00	

(*, $p<0.05$; **, $p<0.01$).

Omission designs

A logical, simple, small but often unsatisfactory design for the diagnosis of nutrient deficiencies is the *omission design*, so-called because with it crop yield is obtained with one treatment for an application of a complete mixture of nutrients and compared with the yields obtained with other treatments for the same mixture but less each nutrient in turn. For example the design in Table 7.7 will give experimental data that can be used to calculate the effects of the omission in turn of each of four nutrients from a complete mixture and may thus be expected to diagnose any deficiency of these four nutrients in an experiment with only five treatments. Similar designs are easily derived for any number of nutrients, that is with $n + 1$ treatments for n nutrients.

Unfortunately this simple type of design can easily produce confusing results because it provides only one comparison per omitted nutrient with the single yield produced by the complete mixture and consequently provides little protection from error effects even when all treatments are replicated. Thus even if the design in Table 7.7 were

Table 7.7 An example omission design of the type not recommended for diagnostic studies on fertilizer requirements.

Row	N	P	K	S
1	1	1	1	1
2	0	1	1	1
3	1	0	1	1
4	1	1	0	1
5	1	1	1	0

replicated three times to provide three comparisons for each of the four nutrients in a 15 plot experiment, it still would provide less protection against error effects to the four comparisons for each nutrient than is provided by the inbuilt replication of the 2^4 factorial (Design 2) for each of the four nutrients in a 16 plot experiment. Moreover it gives no information about interaction effects which unless recognized can cause confusion with the simple yield comparisons. The 2^3 and 2^4 designs of Tables 7.3 and 7.4 are considered minimal for diagnostic studies for these reasons and simpler "non-statistical" designs such as this omission design are not recommended.

7.7 Fertilizer-yield function designs

Designs for experiments that provide data for estimating the square root quadratic yield functions

$$Y = b_0 + b_1 N^{.5} + b_2 N$$

$$Y = b_0 + b_1 N^{.5} + b_2 P^{.5} + b_3 (NP)^{.5} + b_4 N + b_5 P$$

$$Y = b_0 + b_1 N^{.5} + b_2 P^{.5} + b_3 K^{.5} + b_4 (NP)^{.5} + b_5 (NK)^{.5} + b_6 (PK)^{.5} + b_7 N + b_8 P + b_9 K$$

$$Y = b_0 + b_1 N^{.5} + b_2 P^{.5} + b_3 K^{.5} + b_4 S^{.5} + b_5 (NP)^{.5} + b_6 (NK)^{.5} + b_7 (NS)^{.5} + b_8 (PK)^{.5} + b_9 (PS)^{.5} + b_{10} (KS)^{.5} + b_{11} N + b_{12} P + b_{13} K + b_{14} S$$

are chosen on the basis that treatment comparisons for their linear, quadratic and linear × linear polynomial components are most important and consequently should be orthogonal with respect to each other. This basis for the selection of designs is sufficient if it can be safely assumed that the fertilizer-yield relationship is adequately represented by these models so that higher degree effects are negligible and need not be estimated. If this convenient assumption is to be tested however, designs must also be suitable for the estimation of the cubic and linear × quadratic, quadratic × quadratic, linear × linear × linear higher degree interaction effects to demonstrate that they are in fact insignificant and that consequently they and still higher degree effects can be ignored on the basis that they merely represent error aberrations amongst the data. The following designs have been chosen on this basis, so that the important linear, quadratic and linear × linear treatment contrasts are orthogonal with respect to each other and so that these contrasts are also orthogonal with respect to the above higher degree contrasts, or at least nearly so if treatments are evenly spaced (except for Design 7). Any notable correlations involving higher degree contrasts are indicated with each of the designs. The regression procedures for determining the contributions of these higher degree components to regressions are valid in the presence of any such correlations but the accuracy of their determination decreases as the degree of confounding increases, the degree of confounding being indicated by the correlation coefficients.

Design 3 - for one nutrient functions (16 plots, four blocks)

Designs for experiments to estimate square root quadratics for a single nutrient are very simple, requiring only three or four treatment application rates for the single nutrient. The design given in Table 7.8 is for four rates replicated in four blocks giving a total of 16 plots. Because this is the same number of plots as for the next design in Table 7.9 for two nutrient functions, the latter will usually be preferred, if only to check for the unimportance of a second nutrient, unless very small blocks of only four plots must be used. The use of four treatment rates allows the estimation of a cubic effect by *Regression 1* to check that it is not significant and so to justify the use of the quadratic, *Regression 2*, to represent the results of experiments.

The regressions in Table 7.8 can be used to obtain detailed analyses of variance as described above, the sum of squares associated with the underlined coefficients providing sums of squares for the respective treatment effects as listed in the analysis table. The analysis derived from *Regression 1* gives a test of significance for the cubic effect in the data. The second analysis of variance for *Regression 2* assumes that the cubic sum of squares represents only error effects rather than an actual cubic treatment effect so that it can be included with the residue for the estimation of error variance.

Table 7.8 Design 3 for one nutrient yield functions (16 plots, four blocks) and regressions for analysis of variance.

Row	Block 1	Block 2	Block 3	Block 4
	N	N	N	N
1	0	0	0	0
2	1	1	1	1
3	2	2	2	2
4	3	3	3	3

Regression 1 $Y = b_0 + b_1 N^{.5} + b_2 N + \underline{b_3} N^{1.5} + \underline{b_4} L_b + \underline{b_5} Q_b + \underline{b_6} C_b$

Regression 2 $Y = b_0 + b_1 N^{.5} + \underline{b_2} N + \underline{b_3} L_b + \underline{b_4} Q_b + \underline{b_5} C_b$

Regression 3 $Y = a_0 + \underline{a_1} N^{.5} + a_2 L_b + a_3 Q_b + a_4 C_b$

Source of variation		d.f.	Regression 1 Coefficient for sum of squares	d.f.	Regression 2 Coefficient for sum of squares
N effect	Linear	1	a_1	1	a_1
	Quadratic	1	b_2	1	b_2
	Cubic	1	c_3		
Block effect	Linear	1	c_4	1	b_3
	Quadratic	1	c_5	1	b_4
	Cubic	1	c_6	1	b_5
Residue		9		10	
Total		15		15	

Notable confounding: None.

Design 4 - for two nutrient functions (4^2 factorial, 16 plots, two blocks)

Design 4 in Table 7.9 is a standard fractional replicate design with the 16 treatment combinations of four levels of each factor distributed amongst two blocks in such a way that treatment contrasts for the estimation of linear, quadratic, cubic, linear × linear, linear × quadratic polynomial treatment effects and the block effect are all orthogonal

Table 7.9 Design 4 for two nutrient functions (4^2, 16 plots, two blocks) and regressions for analysis of variance.

Row	Block 1	Block 2
	NP	NP
1	01	00
2	02	03
3	10	11
4	13	12
5	20	21
6	23	22
7	31	30
8	32	33

Regression 1 $Y = c_0 + c_1N^{.5} + c_2P^{.5} + c_3(NP)^{.5} + c_4N + c_5P + \underline{c_6}N^{1.5} + \underline{c_7}P^{1.5} + \underline{c_8}N^{.5}P + \underline{c_9}NP^{.5} + \underline{c_{10}}L_b$

Regression 2 $Y = b_0 + b_1N^{.5} + b_2P^{.5} + \underline{b_3}(NP)^{.5} + \underline{b_4}N + \underline{b_5}P + \underline{b_6}L_b$

Regression 3 $Y = a_0 + \underline{a_1}N^{.5} + \underline{a_2}P^{.5} + a_3L_b$

Source of variation	Regression 1		Regression 2	
	d.f.	Coefficient for sum of squares	d.f.	Coefficient for sum of squares
N effects				
Linear	1	a_1	1	a_1
Quadratic	1	b_4	1	b_4
Cubic	1	c_6		
P effects				
Linear	1	a_2	1	a_2
Quadratic	1	b_5	1	b_5
Cubic	1	c_7		
Interactions				
$L_n \times L_p$	1	b_3	1	b_3
$L_n \times Q_p$	1	c_8		
$Q_n \times Q_p$	1	c_9		
Block (L_b)	1	c_{10}	1	b_6
Residue	5		9	
Total	15		15	

Notable confounding : $Q_n \times Q_p$ with L_b, $r = 1.0$.

with respect to each other. The regressions can be used to obtain detailed analyses of variance, the sum of squares associated with the underlined coefficients providing sums of squares for the respective treatment effects as listed in the analysis table. Alternatively F values can be calculated directly from those used for tests of significance for the regression coefficients using the ratios of residual mean squares to that of the main regression as described for Designs 1 and 2. The first analysis in Table 7.9, based on the *Regression 1*, has the purpose of testing the significance of the higher degree trends and hence, if they are small and non-significant, of justifying the use of the square root quadratic model, *Regression 2,* for a fertilizer-yield function. The second analysis is specifically for the *Regression 2*, the sum of squares for the small and non-significant higher degree effects now being included in the residual sum of squares to improve the estimate of error variance and hence the accuracy of the tests of significance.

Example: Two nutrient experiment with Design 4

An experiment with Design 4 is illustrated in Fig. 7.8 with a randomized arrangement of treatments used for its field layout and, in Table 7.10, with the yield data and treatments arranged in the form of a two-way table to facilitate a viewing of treatment and block effects. The treatments 0, 25, 50, 100 kg N/ha and 0, 10, 20, 40 kg P/ha correspond to the design coding and the data produced in each of the two blocks is indicated by putting the block 2 data in square brackets to distinguish them from those of block 1. An inspection of this table suggests that there are positive responses to the N and P treatments and possibly also a positive N × P interaction and a block effect such that the yields in block 2 are greater than those in block 1. The use of the 0:1:2:4 spacing for the treatments in this example produces some small correlations amongst the treatment contrasts and consequently precludes a standard orthogonal analysis of variance but not the analysis with the regressions using the sums of squares associated with the underlined coefficients as described with the previous example. The first analysis of variance based on the *Regressions 1, 2* and *3* shows that 47.64% and 26.92% of the total sum of squares of 1,848,970 is accounted for by linear effects (square root scale) for the N and P applications, giving significant F ratios and also that 6.94% is accounted for by the linear N × linear P interaction and 7.32% by the blocks, as suggested by an inspection of the tabulated data, although neither of the F ratios quite reach the $p<0.05$ level of significance. More importantly however, the analysis shows that the higher degree cubic and linear × quadratic effects are very small and of about the same order of magnitude as the

	Treatment Row	Code	Plot yield
Block 1	7	31	1600
	5	20	1100
	3	10	1040
	1	01	1100
	2	02	950
	4	13	1340
	8	32	1540
	6	23	1900
Block 2	6	22	1760
	2	03	1080
	5	21	1640
	4	12	1740
	7	30	1340
	1	00	990
	3	11	1430
	8	33	2070

Fig. 7.8 Example layout of plots for an experiment with Design 4. Yield data (kg/ha) were produced by the treatments corresponding to the randomizations of the design code in the two blocks.

estimate of error variance provided by the residual mean square (5.62% ÷ 5 = 1.13%), consistent with the expectation that such effects are negligible and mainly represent error aberrations amongst the data. This result justifies the use of the *Regression 2* to represent the relationship and the use of the second analysis of variance based on *Regressions 2* and *3* with the higher degree effects incorporated into the residual sum of squares to obtain a residual mean square (6.52% ÷ 9 = 0.72%) with 9 degrees of freedom. The second analysis with this residual mean square gives more sensitive tests of significance resulting in higher significance ratings for the F ratios and in

Table 7.10 Tabulated data and analyses of variance for the example experiment of Fig. 7.8. Block 2 data in square brackets.

kg P/ha	kg N/ha 0	25	50	100
0	[990]	1040	1100	[1340]
10	1100	[1430]	[1640]	1600
20	950	[1740]	[1760]	1540
40	[1080]	1340	1900	[2070]

Regression 1 $Y = 910.4 - 80.45N^{.5} + 201.3P^{.5} + 2.390(NP)^{.5} + 33.77N + 47.99P$
$- \underline{2.236N^{1.5}} + \underline{2.748P^{1.5}} + \underline{2.433N^{.5}P} - \underline{0.6826NP^{.5}} + \underline{93.85L_b}$

Regression 2 $Y = 934.3 + 63.26N^{.5} + 61.97P^{.5} + \underline{10.65}(NP)^{.5} - \underline{3.744N} - \underline{7.259P}$
$+ \underline{90.69L_b}$

Regression 3 $Y = 792.0 + \underline{64.32N^{.5}} + \underline{76.47P^{.5}} + 90.69L_b$

Source of variation	From regression 1 d.f.	% Sum of squares	F ratio	From regression 2 d.f.	% Sum of squares	F ratio
N main effects						
Linear	1	47.64	42.40**	1	47.64	50.18***
Quadratic	1	1.59	1.41	1	1.59	1.67
Cubic	1	1.11	0.99			
P main effects						
Linear	1	26.92	23.96**	1	26.92	28.36***
Quadratic	1	0.96	0.85	1	0.96	1.01
Cubic	1	0.11	0.10			
Interactions						
$L_n \times L_p$	1	6.94	6.18	1	6.94	7.31*
$L_n \times Q_p$	1	1.40	1.25			
$Q_n \times L_p$	1	0.28	0.25			
Block effect						
Linear (L_b)	1	7.32	6.52	1	7.12	7.50*
Residue	5	5.62		9	8.54	
Total (%)	15	100.00		15	100.00	
Actual total		1848970			1848970	

(*, $p<0.05$; **, $p<0.01$; ***, $p<0.001$).

particular for the interaction and block effects, now rated $p<0.05$. The two analyses of variance thus support the use of the square root quadratic model for a fertilizer-yield function as well as supporting the effects suggested by an inspection of the data.

Design 5 - for three nutrient quadratic yield functions (3^3 factorial, 27 plots, three blocks)

Design 5 (Table 7.11) is a standard and very efficient design for estimating quadratic yield functions for three nutrients, requiring only 27 plots distributed amongst three small blocks, each of only nine plots. If treatments are evenly spaced on the square root scale, then all linear, quadratic and interaction contrasts are orthogonal with respect to each other. Since however the design has only three levels for each nutrient it cannot be used to test the assumed insignificance of cubic or higher degree effects. Accordingly the design is suitable for situations where previous experience has shown that cubic effects are likely to be negligible.

Design 6 - for three nutrient yield functions (4^3 ½ factorial, 32 plots, two blocks)

Design 6 (Table 7.12) is an alternative to Design 5 with four levels of each nutrient so that cubic effects can be estimated to check on the adequacy of quadratic models for the representation of the relationship. The important linear, quadratic, cubic and linear × linear interaction treatment contrasts are orthogonal with respect to each other for evenly spaced treatment rates. There is however a confounding of the quadratic N with the quadratic P × quadratic K interaction contrast ($r = 1.0$) and of the linear × quadratic contrasts with each other ($r = 0.80$). This confounding is unimportant if these higher degree interaction effects are insignificant.

The analysis of variance procedure indicated in Table 7.12 is only for linear and quadratic effects. Since the nutrients are applied with four treatment rates, cubic effects can be determined with a corresponding procedure to that described above, using the sums of squares associated with the underlined coefficients in the regression

$$Y = c_0 + c_1N^{.5} + c_2P^{.5} + c_3K^{.5} + c_4(NP)^{.5} + c_5(NK)^{.5} + c_6(PK)^{.5}$$
$$+ c_7N + c_8P + c_9K + \underline{c_{10}}N^{1.5} + \underline{c_{11}}P^{1.5} + \underline{c_{12}}K^{1.5} + \underline{c_{13}}L_b$$

Table 7.11 Design 5 for three nutrient quadratic yield functions.(3^3, 27 plots, three blocks) and regressions for analysis of variance.

Row	Block 1	Block 2	Block 3
	NPK	NPK	NPK
1	000	001	002
2	011	012	010
3	022	020	021
4	101	102	100
5	112	110	111
6	120	121	122
7	202	200	201
8	210	211	212
9	221	222	220

Regression 1 $Y = b_0 + b_1N^{.5} + b_2P^{.5} + b_3K^{.5} + \underline{b_4}(NP)^{.5} + \underline{b_5}(NK)^{.5} + \underline{b_6}(PK)^{.5} + \underline{b_7}N + \underline{b_8}P + \underline{b_9}K + \underline{b_{10}}L_b + \underline{b_{11}}Q_b$

Regression 2 $Y = a_0 + \underline{a_1}N^{.5} + \underline{a_2}P^{.5} + \underline{a_3}K^{.5} + a_4L_b + a_5Q_b$

Source of variation		d.f.	Coefficient for sum of squares
N effects	Linear	1	a_1
	Quadratic	1	b_7
P effects	Linear	1	a_2
	Quadratic	1	b_8
K effects	Linear		a_3
	Quadratic		b_9
Interactions	$L_n \times L_p$	1	b_4
	$L_n \times L_k$	1	b_5
	$L_p \times L_k$	1	b_6
Block effects	L_b	1	b_{10}
	Q_b	1	b_{11}
Residue		15	
Total		26	

Notable confounding (correlations): None.

Table 7.12 Design 6 for three nutrient yield functions (4^3, 32 plots, two blocks) and regressions for analysis of variance.

	Block 1		Block 2	
Row	NPK	NPK	NPK	NPK
1	000	202	003	201
2	011	213	012	210
3	022	220	021	223
4	033	231	030	232
5	101	303	102	300
6	110	312	113	311
7	123	321	120	322
8	132	330	131	333

Regression 1 $Y = b_0 + b_1N^{.5} + b_2P^{.5} + b_3K^{.5} + \underline{b_4}(NP)^{.5} + \underline{b_5}(NK)^{.5}$

$$+ \underline{b_6}(PK)^{.5} + \underline{b_7}N + \underline{b_8}P + \underline{b_9}K + \underline{b_{10}}L_b$$

Regression 2 $Y = a_0 + \underline{a_1}N^{.5} + \underline{a_2}P^{.5} + \underline{a_3}K^{.5} + a_4L_b$

Source of variation		d.f.	Coefficient for sum of squares
N effects	Linear	1	a_1
	Quadratic	1	b_7
P effects	Linear	1	a_2
	Quadratic	1	b_8
K effects	Linear	1	a_3
	Quadratic	1	b_9
Interactions	$L_n \times L_p$	1	b_4
	$L_n \times L_k$	1	b_5
	$L_p \times L_k$	1	b_6
Block effect	Linear	1	b_{10}
Residue		21	(from regression 1)
Total		31	

Notable confounding (correlations): Q_n with $Q_p \times Q_k$, r = 1.00; $Q_p \times L_p$ with $L_p \times Q_k$, r = 0.80.

Design 7 - for four nutrient yield functions (4^4 ¼ factorial, 64 plots, four blocks)

Experiments with four levels of four nutrients represent about the limit of practicality for fertilizer field experiments. Design 7 (Table 7.13) is of a practicable size with relatively small blocks because it is a ¼ replicate design. The fractional replication only produces confounding with higher degree interaction contrasts, the important linear, quadratic, cubic and linear × linear contrasts all being orthogonal with respect to each other.

The analysis of variance indicated in Table 7.13 does not include cubic effects for the nutrient treatments. These can be determined by extending the first regression to include $N^{1.5}$, $P^{1.5}$, ... variables.

Central composite "response surface" designs

The above factorial designs have been selected for the estimation of fertilizer-yield functions which may also be called response surfaces. In standard references (for example Box and Wilson 1951; Cochran and Cox 1957; Box and Draper 1959; Hill and Hunter 1966; John and Quenouille 1977; Khuri and Cornell 1987) another type of design is usually described for the estimation of response surfaces, called central composite designs, and experimenters have consequently often been led (or misled) into attempting to use these for the estimation of fertilizer-yield functions with fertilizer experiments. In these designs the treatment combinations are centred around an anticipated optimum and chosen in such a way that linear, quadratic and interaction contrasts are orthogonal. The designs are excellent if the estimate of the central combination of treatments is located reasonably close to the actual optimum, as is usually possible for example with experiments for industrial manufacturing situations. They are however usually unsatisfactory for fertilizer experiments because unexpected seasonal and site conditions can produce an optimum very different to that anticipated. For example the optimal rate for one of the nutrients being studied may prove to be nil under the experimental conditions so that estimates of optimal rates for all of the nutrients must be based on extrapolations beyond the range of the treatment *combinations* for a central composite design with consequent extrapolation errors. Such extrapolation effects can produce quite unrealistic forms in estimates of response surfaces within the ranges of the treatment rates considered individually as illustrated by Colwell and Stackhouse (1970). Consequently optimal rates that are calculated from response surfaces that have can been estimated from data obtained from experiments with

Table 7.13 Design 7 for four nutrient yield functions (4^4, 64 plots, four blocks) and regressions for analysis of variance.

Row	Block 1	Block 2	Block 3	Block 4
	NPKS	NPKS	NPKS	NPKS
1	0000	0032	0013	0021
2	0131	0103	0122	0110
3	0212	0220	0201	0233
4	0323	0311	0330	0302
5	1011	1023	1002	1030
6	1120	1112	1133	1101
7	1203	1231	1210	1222
8	1332	1300	1321	1313
9	2022	2010	2031	2003
10	2113	2121	2100	2132
11	2230	2202	2223	2211
12	2301	2333	2312	2320
13	3033	3001	3020	3012
14	3102	3130	3111	3123
15	3221	3213	3232	3200
16	3310	3322	3303	3313

Regression 1

$$Y = b_0 + b_1N^{.5} + b_2P^{.5} + b_3K^{.5} + b_4S^{.5} + \underline{b_5}(NP)^{.5} + \underline{b_6}(NK)^{.5} + \underline{b_7}(NS)^{.5} + \underline{b_8}(PK)^{.5}$$

$$+ \underline{b_9}(PS)^{.5} + \underline{b_{10}}(KS)^{.5} + \underline{b_{11}}N + \underline{b_{12}}P + \underline{b_{13}}K + \underline{b_{14}}S + \underline{b_{15}}L_b + \underline{b_{16}}Q_b + \underline{b_{17}}C_b$$

Regression 2

$$Y = a_0 + \underline{a_1}N^{.5} + \underline{a_2}P^{.5} + \underline{a_3}K^{.5} + \underline{a_4}S^{.5} + a_5L_b + a_6Q_b + a_7C_b$$

Source of variation		d.f.	Coefficient for sum of squares
N effects	Linear	1	a_1
	Quadratic	1	b_{11}
P effects	Linear	1	a_2
	Quadratic	1	b_{12}
K effects	Linear	1	a_3
	Quadratic	1	b_{13}
S effects	Linear	1	a_4
	Quadratic	1	b_{14}

Table 7.13 continued.

Interaction effects		
$L_n \times L_p$	1	b_5
$L_n \times L_k$	1	b_6
$L_n \times L_s$	1	b_7
$L_p \times L_k$	1	b_8
$L_p \times L_s$	1	b_9
$L_k \times L_s$	1	b_{10}
Block effects		
L_b	1	b_{15}
Q_b	1	b_{16}
C_b	1	b_{17}
Residue	46	(regression 1)
Total	63	

Notable confounding: $L_n \times L_s$ with $Q_p \times Q_k$, $r = 0.40$; $L_n \times Q_p$ with $Q_k \times L_s$, $r = 0.40$; $L_n \times L_p$ with $L_k \times Q_s$, $r = -0.36$; $L_k \times L_s$ with $Q_n \times L_p$, $r = -0.18$.

central composite designs can be very misleading, even though the estimated rates are within the treatment ranges and so are not obvious as estimates that have been based on extrapolations, that is on extrapolations beyond the range of the treatment combinations. For this reason central composite type designs are *not* recommended for fertilizer experiments.

7.8 Fertilizing programs

Programs for successive crops

The series of fertilizer applications that are made to the successive crops of a farming system can be described as a *fertilizing program* for that system and a program that produces the maximum economic return for the system can be described as being an *optimal fertilizing program*. If each fertilizer application affects only the crop to which it is applied, with no effects of fertilizer residues on following crops, then an optimal program is obtained simply by determining the optimal rate for each of the successive crops, using the yield function designs already described. Thus for a region with an established and stable cropping system, a fertilizing program can be envisaged as consisting simply of the continuing series of fertilizer applications at optimal rates that replenishes the soil nutrients removed with each crop and maintains the soil nutrient levels at an optimal level. In this situation experiments are needed only to determine the optimal rates for the crops, each considered individually. When however initial soil nutrient levels are low so that optimal rates are greater than such maintenance rates, residues of the successive fertilizer applications remain in the soil increasing soil nutrient levels and thus reducing the fertilizer requirements of the successive crops. For example when an agricultural system is being established in a new region in which there is a severe soil nutrient deficiency, a fertilizing program may be envisaged as consisting of a series of decreasing fertilizer application rates, commencing with a high rate and followed by progressively lower rates, all converging towards a maintenance rate for a stable agricultural system with a higher soil nutrient level. Similarly when a perennial crop such as an orchard of fruit trees is being established, an optimal program may be envisaged as consisting initially of a series of increasing application rates to meet the increasing nutrient requirements of the developing crop followed by a series of decreasing rates, converging towards a regular maintenance rate for the established crop. Corresponding simultaneous programs can be envisaged for two or more nutrient applications for situations where there are multiple nutrient deficiencies, with different programs for each nutrient as determined by differing effects of the residues on the level of each nutrient in the soil. The optimal fertilizing program in such situations produces the maximum economic return as described in section 3.9.

Experiments for estimating optimal fertilizing programs are obviously much more difficult than those for estimating optimal rates for individual crops and this undoubtedly explains why there seem to have been few attempts to estimate optimal fertilizing programs with fertilizer experiments. Firstly the experiments must be long-term, to cover the effects of residues from previous fertilizer applications as well as of fresh applications on a succession of crops, and this may entail difficult management problems for the duration of the experiment. Then, after the completion of experiments that have lasted perhaps several years, the experimental results may be difficult to interpret because they have been affected, possibly drastically, by the sequence of seasonal growing conditions that happened to prevail during the course of the experiment. Examples of these difficulties with programs for a single fertilizer nutrient, in Australia and Brazil, are given by Colwell (1985a, b). For these reasons the following designs are given rather tentatively, for use in special situations where, for example, variations between growing conditions for successive crops are likely to be small or where programs can be compressed into single years and replicated in successive years, as for some multiple cropping systems. Corresponding designs for the much more complex situation of programs for multiple nutrient deficiencies are not attempted.

Data

Treatments for fertilizing programs consist of alternative sets of fertilizer application rates for a sequence of crops, not necessarily of the same species, and data for the effects of the treatment programs must be determined by combining data for the effects of each application in the program on the individual crops of the sequence. Moreover if the data are to be used to estimate an economically optimal program, allowance must be made for the time scale in which the successive economic returns are received. For these reasons the accumulated yield data are conveniently expressed in the form of accumulated time discounted profits (equation 3.21) rather than accumulated crop yields. For example if programs of five successive fertilizer applications were applied to three successive annual crops of soybeans followed by two annual crops of rice, data for the treatment effects might be obtained by calculating the total time discounted return $\Sigma\Pi_D$ from the five crops, discounted to the time of the first purchase of fertilizer. Thus assuming an interest conversion period of one year between each purchase and each return with sale of the crops,

$$\Sigma\Pi_D = \frac{V_sY_1}{(1+R)} - CN_1 + \frac{V_sY_2}{(1+R)^2} - \frac{CN_2}{(1+R)} + \frac{V_sY_3}{(1+R)^3} - \frac{CN_3}{(1+R)^2} + \frac{V_rY_4}{(1+R)^4} - \frac{CN_4}{(1+R)^3} + \frac{V_rY_5}{(1+R)^5} - \frac{CN_5}{(1+R)^4} - \sum Q \tag{7.10}$$

where V_s and V_r are the values per unit of soybeans and rice, Y_1, ... , Y_5 , are the successive crop yields, R is interest rate, C is the cost per unit of the nutrient N and N_1 , ... , N_5 are the successive nutrient rates of the fertilizing program. The optimal fertilizing program is that sequence of rates N_1 , ... , N_5 that produces the maximum profit $\Sigma\Pi_D$. Similar profit functions can be derived for other types of fertilizing program, based similarly on the discounted profit equation (3.21) and in much more complex forms, for multiple nutrient fertilizing programs. The main problem is a practical one, to obtain reliable estimates of the successive yield functions, rather than the mathematical exercise of formulating and applying profit equations.

A factorial basis for program treatments

The many possible combinations of rates that are possible in fertilizing programs can produce confusion both in the planning of experiments and the interpretation of the data they produce. To assist in this respect treatment sequences can be chosen on the two bases illustrated by the example treatment programs in Table 7.14. Thus for the example the programs 1, 2 and 3 apply different amounts of fertilizer but all on the same basis of the relative sequence of rates defined by the ratios 20 : 10 : 5 : 3 : 2. Such a set of treatments can be expected to indicate the optimal rate for the constant relative sequence. Similarly the programs 4, 5 and 6 apply the same total amount of fertilizer in a five year period but with different relative sequences of application and such a set of treatments can be expected to indicate optimal sequences for a particular total application of fertilizer. Factorial combinations of such relative sequences and rates of application can be used to derive an orderly series of program treatments for the estimation of an optimal program, as in Table 7.15.

Data obtained from such an experiment with replications may be analysed by the usual procedure for a randomised block design, profits ($\Sigma\Pi_D$) can be compared for the alternative sets of ratios with the same total fertilizer application and optimal nutrient applications for any particular set of ratios can be calculated from the equation

$$\Sigma\Pi_D = b_0 + b_1 T^{.5} + b_2 T \qquad (7.11)$$

by $d(\Sigma\Pi_D)/dT = 0$ where T is the total nutrient application. Thus if for example an experiment with the Table 7.15 treatments produced data such that the ratios 11 : 5 : 2 : 1 : 1 gave consistently highest values $\Sigma\Pi_D$ for each of the different total applications and if the calculated optimal rate from the regression equation for (7.11) was T = 260 for this set of ratios, then this result would indicate the optimal program 143, 65, 26, 13, 13 kg nutrient/ha.

Table 7.14 Example sets of fertilizer programs for five successive crops.

Treatment	Crop					Total
	1	2	3	4	5	
Constant ratio						
Program 1	50	25	12	8	5	100
Program 2	100	50	25	15	10	200
Program 3	200	100	50	30	20	400
Constant total						
Program 4	100	50	24	16	10	200
Program 5	110	50	20	10	10	200
Program 6	160	10	10	10	10	200

Table 7.15 Example design derived by factorial combination of ratios for successive fertilizer rates and total nutrient applied for a five crop fertilizing program.

Program	Fertilizer treatment rates						
	Crop					Total	Ratios
	1	2	3	4	5	fertilizer	
1	50	25	12	8	5	100	10:5:2.4:1.6:1
2	100	50	24	16	10	200	"
3	200	100	48	32	20	400	"
4	55	25	10	5	5	100	11:5:2:1:1
5	110	50	20	10	10	200	"
6	220	100	40	20	20	400	"
7	80	5	5	5	5	100	16:1:1:1:1
8	160	10	10	10	10	200	"
9	320	20	20	20	20	400	"

Serial design

Factorial designs can be used to derive serial designs, as described rather critically by John and Quenouille (1977), in which treatments consist of applications of the same nutrient to successive crops. Thus the design in Table 7.16 is derived from the factorial Design 5 in Table 7.11 with treatments for applications of *three different nutrients to one crop* replaced by treatments for applications of *one nutrient to three different (successive) crops*. The experiment plots receive the

Table 7.16 Example serial design for application of a nutrient at rates 10, 40 and 90 kg/ha to three successive crops.

Block 1			Block 2			Block 3		
N_1	N_2	N_3	N_1	N_2	N_3	N_1	N_2	N_3
Crop 1	Crop 2	Crop 3	Crop 1	Crop 2	Crop 3	Crop 1	Crop 2	Crop 3
10	10	10	40	10	10	90	10	10
40	40	10	90	40	10	10	40	10
90	90	10	10	90	10	40	90	10
40	10	40	90	10	40	10	10	40
90	40	40	10	40	40	40	40	40
10	90	40	40	90	40	90	90	40
90	10	90	10	10	90	40	10	90
10	40	90	40	40	90	90	40	90
40	90	90	90	90	90	10	90	90

applications listed in the first column of each block for a first crop, then the same plots receive the applications in the second column for a following, second crop and finally the same plots receive the applications in the third column for a third crop. Denoting the yield data for the three crops by Y_1, Y_2, Y_3 and the nutrient treatment rates for these three crops by N_1, N_2, N_3 the results for each of the successive crops in the experiment can be represented by yield functions with the forms

$$Y_1 = b_0 + b_1 N_1^{.5} + b_2 N_1 + b_3 L_b + b_4 Q_b$$

$$Y_2 = b_0 + b_1 N_1^{.5} + b_2 N_2^{.5} + b_3 (N_1 N_2)^{.5} + b_4 N_1 + b_5 N_2 + b_6 L_b + b_7 Q_b$$

$$Y_3 = b_0 + b_1 N_1^{.5} + b_2 N_2^{.5} + b_3 N_3^{.5} + b_4 (N_1 N_2)^{.5} + b_5 (N_1 N_3)^{.5} + b_6 (N_2 N_3)^{.5}$$
$$+ b_7 N_1 + b_8 N_2 + b_9 N_3 + b_{10} L_b + b_{11} Q_b \quad (7.12)$$

where L_b and Q_b are dummy block variables. The interaction terms $(N_1 N_2)^{.5}$, etc. in these regressions are expected to be negative, representing substitution effects from previous applications of the nutrient. For example the rate N_2 required to produce a particular yield level is expected to decrease with increase in the previous application rate N_1 giving a negative interaction effect.

Regression estimates of these functions can be used to calculate values for N_1, N_2 and N_3 producing maximum yields or maximum profits, with an appropriate time discounting of costs, for each of the successive crops. More attractively, the yield functions for the three crops can be combined to calculate the total time discounted total profit received at the end of the three year period, as by

$$\Sigma\Pi = V\left[Y_1(1+R)^2 + Y_2(1+R) + Y_3\right]$$
$$- C_n\left[N_1(1+R)^3 + N_2(1+R)^2 + N_3(1+R)\right] \quad (7.13)$$

for an interest conversion period of one year, ignoring fixed costs and omitting block effects to obtain an experiment site average. The profit function can then be used to calculate values for the variables N_1, N_2 and N_3 that produce the maximum profit $\Sigma\Pi$ and hence a direct estimate of the optimal fertilizing program for the three successive crops. The procedure may be unsatisfactory, however, for crops in successive years because:

1. Experiments for more than three or four years become large and unwieldy.
2. The data produced for each successive crop can be greatly influenced by the growing conditions that occur in the successive years.

Thus for example if a profit function of form (7.13) is estimated from data for which the growing conditions were bad for the first crop and good for the third crop, the calculated values for the optimal rates may have values such that $N_1 < N_2 < N_3$ rather than the normally expected sequence $N_1 > N_2 > N_3$. Examples of such seasonal effects on estimates of optimal programs are given in Colwell (1985a,b). These problems will not apply however if the experiments are for a multiple cropping system in which the successive crops are grown in the successive seasons of individual years.

In general then, although standard designs can be adapted for experiments with fertilizing programs for successive crops and mathematical procedures employed to calculate estimates of optimal programs, results are only likely to be satisfactory if they are not greatly affected by variations in the growing conditions for each of the successive crops. Thus they are better suited to studies for multiple cropping systems with the successive crops in individual years rather than to long term experiments covering several successive years.

Programs for individual crops

Serial type designs can also be used for the estimation of optimal programs of successive applications of fertilizer to individual crops. For example N fertilizer may be applied to a maize crop at two stages of growth, as at the time of sowing and of tillering some time later, using *Design 4* and the analysis of variance procedure of Table 7.9 and substituting the treatment rates N_s and N_t for N and P. Thus with the treatment rates N_s = 4, 22, 54, 100 kg N/ha for fertilizer applied at sowing and N_t = 0, 10, 20, 40 kg N/ha for fertilizer applied at tillering the design in Table 7.17 would be used to obtain data to estimate the function

$$Y = b_0 + b_1 N_s^{.5} + b_2 N_t^{.5} + b_3 (N_s N_t)^{.5} + b_4 N_s + b_5 N_t + b_6 L_b \qquad (7.14)$$

Optimal rates N_s and N_t for the program of two applications of N to the crop can then be calculated by the optimal rate equations (3.10), substituting again N_s and N_t for N and P as described in chapter 3.

Table 7.17 Example serial design derived from Design 4 with treatments for applications of N fertilizer at sowing and tillering, N_s = 4, 22, 54, 100 and N_t = 0, 10, 20, 40 kg N/ha.

Block 1			Block 2		
Code	N_s	N_t	Code	N_s	N_t
01	4	10	00	4	0
02	4	20	03	4	40
10	22	0	11	22	10
13	22	40	12	22	20
20	54	0	21	54	10
23	54	40	22	54	20
31	100	10	30	100	0
32	100	20	33	100	40

Chapter 8
General Soil Fertility Models

8.1 Nature of general soil fertility models

The effects of treatments in experiments vary with the conditions under which the experiments are carried out so that strictly, the results of any particular experiment apply only to the set of conditions that prevailed for that experiment and different, though possibly similar, results must be expected from experiments with other conditions. This presents an obvious problem with the application of results obtained with fertilizer experiments because conditions commonly vary widely in agricultural regions so that the results of individual experiments, and the estimates of fertilizer requirements they provide, are likely to be misleading if used directly as a guide for fertilizer use under other conditions, at other sites and in other years. In particular the results of experiments that have been carefully carried out at carefully selected sites or, in other words, under particularly favourable growing conditions, may be very misleading if applied directly as a guide for fertilizer use by farmers. An unfortunate consequence is a general scepticism amongst farmers towards the results and recommendations produced by scientists, especially in developing countries where the differences between conditions on a research establishment and those in farmers' fields is often only too obvious. Of course experiments should be carefully carried out, at carefully selected sites but if the results they produce are to be used for the benefit of farmers, they must be adjusted to allow for the conditions in farmers' fields, at other sites and in other years. *General soil fertility models* are intended to provide a scientific and statistical basis for such adjustments by representing the relationships between the results of fertilizer experiments and site variables in such a way that corresponding results can be estimated for the particular sets of growing conditions, or at least for the expected sets of growing conditions, that will apply for sites in the region represented by the experiments. A statistical procedure for developing such models is described in this chapter whereby the results of appropriately designed fertilizer experiments are quantified in the form of special variables, called *yield variables*, and regression relationships are established between these yield variables and *site variables* in such a way that the relationships can be used collectively to predict corresponding results for nominated values of the site variables. In so far as the results of fertilizer experiments can be regarded as constituting a sampling of the soil fertility of a region, the models can be regarded as representing the soil fertility and the factors that affect it, in a variable region. The set of regressions that constitute the model have been called a *general soil fertility model* on this basis (Colwell *et al.* 1988).

The development of a general soil fertility model for a region involves carrying out a series of fertilizer experiments to obtain data for yield variables and site variables that represent the range of soil fertility in the region and then establishing a set of regression relationships of the yield variables on site variables that will collectively constitute the model. To obtain consistent sets of data for these regressions the yield variable data must be obtained from a series of experiments, all with the same design and treatments (but different randomizations), just as more obviously the site variable data for each experiment must be obtained by the same procedures. In addition, for reasons that will be explained in section 8.6, the yield variables must represent orthogonal components of the yield variance for each experiment so that the regressions of the model can be used collectively to make predictions. This latter requirement is met by making the yield variables correspond to the statistically independent (orthogonal) components of an analysis of variance of the experimental data. In the procedure to be described, factorial designs are used for the experiments and yield variables are derived to correspond to the orthogonal polynomial components of a detailed analysis of variance of the data from these experiments. The site variables are defined more simply as those that affect the results of the experiments, such as those that represent soil composition, kind of soil, weather and farming practices at each experiment site. Accordingly the descriptions that follow are mainly of statistical procedures for (i) the computation of yield variables to represent the results of fertilizer experiments with factorial designs, (ii) the establishment of regression relationships between these yield variables and site variables and (iii) the use of these regressions to predict the results of fertilizer experiments and hence fertilizer requirements from nominated values for site variables. The scientist has the responsibility of choosing appropriate site variables and regression forms for the general models for particular regions.

The procedure for developing and using general soil fertility models is conveniently introduced using the data in Table 8.1 for a series of seven fertilizer experiments which show an obvious and simple relationship between the levels of yield and yield response to P fertilizer treatment rates in the experiments and the single site variable T, where T is a standard "soil test" analysis for phosphorus on surface soil samples taken from the experiment sites before the application of the treatments. The data, although based on actual experiments (Colwell *et al.* 1988), have been selected and modified for this introduction. Thus the data show a convenient range of yield responses to the P fertilizer as illustrated in Fig. 8.1 and an obvious error effect in the data of experiment C. In the real world many more experiments, perhaps with several fertilizer nutrients, are required to obtain an adequate representation of a region, error effects are often not obvious and many site variables are required to adequately explain the variations amongst the results for the different experiments.

8.2 Yield variables

Computation with regressions

Many values can be calculated from the data of fertilizer experiments to represent particular aspects of soil fertility such as the maximum yield attainable with fertilizer or maximum yield response to fertilizer application, and these values can be used to estimate regression relationships with site variables such as soil analyses. As already mentioned however, if such regressions are to be used jointly to estimate fertilizer-yield functions, they must be for orthogonal components of the variance of the experimental data. The trend coefficients of orthogonal polynomial regressions for the fertilizer-yield relationship that have been estimated from experimental data, such as the p_0, p_1 and p_2 of the single nutrient function $Y = p_0Z + p_1L + p_2Q$, relate directly to the components of an orthogonal analysis of variance. Consequently they may be used as dependent variables for a set of regressions on site variables and also consequently, the regressions can then be used collectively to estimate the coefficients of corresponding functions for particular values of the site variables. The functions

Table 8.1 Replicate means of crop yields and site variable values for soil phosphorus (T), for seven fertilizer experiments.

Experiment	Plot yield (kg/ha)				Site variable
	Treatment (kg P/ha)				
	0	15	30	60	T (ppm P)
A	80	1500	1704	2760	1
B	980	2180	2350	2890	2
C	1690	1780	2340	2990	3
D	1850	2500	2700	2480	5
E	2500	2890	2870	2950	12
F	2960	3020	2970	2990	24
G	3380	3300	3400	3100	78

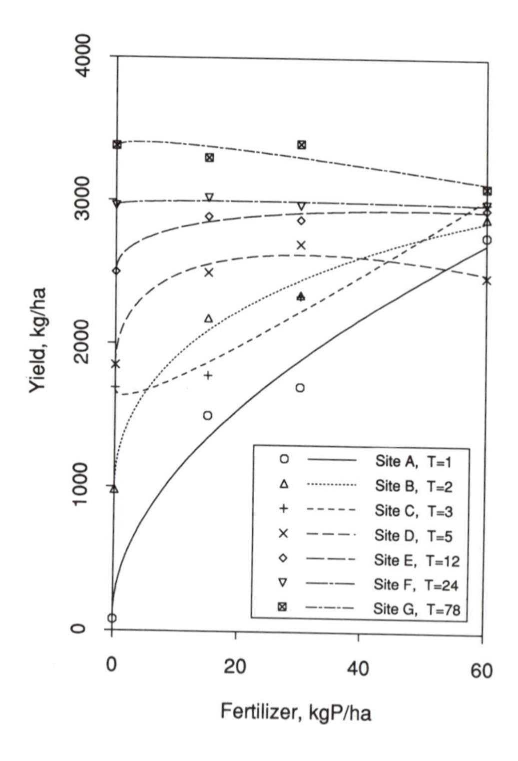

Fig. 8.1 Graphs for estimates of regressions with the model form $Y = b_0 + b_1P^{.5} + b_2P$ with points for the yield data of Table 8.1.

estimated in this way correspond in effect to estimates of the functions that would be obtained by carrying out fertilizer experiments at sites with the site variable values. The estimated functions can consequently be used to calculate yields, optimal rates, etc. in the same way as if they had been estimated directly with experiments. This type of procedure as described originally by Colwell (1967, 1968) can be applied more conveniently using coefficients of ordinary non-orthogonal polynomial regressions that relate directly to these orthogonal trend coefficients, differing only by constant factors. The selected regression coefficients can be used as *yield variables* for general models because they represent independent or orthogonal trends in the data for each experiment which in turn correspond with data features that may be expected to relate to site variables. Thus the values of the selected coefficients of the regressions estimated from the data for each experiment become data values for the yield variables that are used for regressions on site variables.

The derivation of yield variable data for estimating regressions for a general model, from regressions that have been estimated from the data of fertilizer experiments, is illustrated with the example data in Table 8.1. Thus if the example data were obtained from experiments with nutrient treatment rates P and Design 3 of Table 7.8, the values of estimates for the underlined coefficients in the regressions

$$Y = c_0 + c_1 P^{\cdot 5} + c_2 P + \underline{c_3} P^{1\cdot 5} + c_4 L_b + c_5 Q_b + c_6 C_b \tag{8.1}$$

$$Y = b_0 + b_1 P^{\cdot 5} + \underline{b_2} P + b_3 L_b + b_4 Q_b + b_5 C_b \tag{8.2}$$

$$Y = a_0 + a_1 P^{\cdot 5} + a_2 L_b + a_3 Q_b + a_4 C_b \tag{8.3}$$

provide data for yield variables since these particular coefficients relate directly to orthogonal components of the analyses of variance of the experiment data. The coefficients for the block variables L_b, Q_b and C_b also correspond to orthogonal components of the analysis but since these only represent trends within experiment sites and have a mean value of zero (section 7.3), they are irrelevant for general model relationships with site variables. Also since the cubic regression (8.1) is used simply to check that the cubic trend is insignificant and hence to justify the use of the quadratic regression form (8.2) as explained in chapter 7, the coefficient c_3 can be expected not to be needed for general models. For this example regression estimates of the coefficients b_2 and a_1 thus provide data for yield variables as listed in the column B_p in Table 8.2 and A_p in Table 8.3. Values for the coefficient a_1 serve as an overall measure of yield response to the fertilizer treatments in each experiment and values for the coefficient

b_2 serve as an overall measure of the curvature in the quadratic yield function not provided for by a_1, as will be illustrated below in section 8.3. The additional yield variable $\overline{Y}$ for the mean of the yield data, corresponding to the zero orthogonal polynomial trend in analyses of variance, is also required as a measure of level of yield and for this reason is given in Table 8.4 as a zero degree "regression" equation, to complete the series of regressions that provide data values for the yield variables A_p, B_p and $\overline{Y}$.

Table 8.2 Regression estimates of the yield-fertilizer relationship $Y = b_0 + b_1P^{.5} + b_2P$ from the data in Table 8.1. Estimates for the coefficient b_2 provide data values for the yield variable B_p.

Experiment	Regression	B_p
A	$Y = 102.9 + 304.8P^{.5} + 4.013P$	4.013
B	$Y = 992.5 + 328.2P^{.5} - 11.25P$	−11.25
C	$Y = 1678.8 - 73.74P^{.5} + 31.94P$	31.94
D	$Y = 1841.4 + 285.4P^{.5} - 25.85P$	−25.85
E	$Y = 2505.7 + 122.5P^{.5} - 8.648P$	−8.648
F	$Y = 2963.3 + 15.25P^{.5} - 1.656P$	−1.656
G	$Y = 3386.6 + 35.99P^{.5} - 8.663P$	−8.663

Table 8.3 Regressions for the yield-fertilizer relationship $Y = a_0 + a_1P^{.5}$ estimated from the data in Table 8.1. Estimates for the coefficient a_1 provide data values for the yield variable A_p.

Experiment	Regression	A_p
A	$Y = 80.19 + 334.8P^{.5}$	334.8
B	$Y = 1056 + 244.2P^{.5}$	244.2
C	$Y = 1496 + 164.7P^{.5}$	164.7
D	$Y = 1988 + 92.37P^{.5}$	92.37
E	$Y = 2555 + 57.97P^{.5}$	57.97
F	$Y = 2973 + 2.885P^{.5}$	2.885
G	$Y = 3418 - 28.69P^{.5}$	−28.69

Table 8.4 Equations for the mean yield $Y = \overline{Y}$ estimated from the data in Table 8.1, giving values for the yield variable $\overline{Y}$.

Experiment	Equation	$\overline{Y}$
A	Y = 1511.0	1511.0
B	Y = 2100.0	2100.0
C	Y = 2200.0	2200.0
D	Y = 2382.5	2382.5
E	Y = 2802.5	2802.5
F	Y = 2985.0	2985.0
G	Y = 3295.0	3295.0

Recommended computational procedure

An alternative, more general and more efficient procedure for obtaining the same yield variable data from experiment data is to calculate all of the values from the quadratic regression. This procedure is more efficient because the equations for the calculation of the lower degree coefficients and the mean for the yield variable data are required at a later stage, to calculate estimates of yield functions from the regressions of the yield variables on site data.

Given a regression estimate of the fertilizer-yield function for the nutrient N

$$Y = b_0 + b_1 N^{.5} + b_2 N$$

estimated from a set of data with the n treatment rates N_i, i = 1, 2, ... , n, the coefficients b_0, b_1 and b_2 can be used to calculate the value of the coefficient a_1 in the regression

$$Y = a_0 + a_1 N^{.5}$$

and the value of the mean yield, $\overline{Y}$, each estimated from the same data. Thus

$$a_1 = b_1 + b_2 h_n \tag{8.4}$$

and

$$\overline{Y} = b_0 + b_1 \overline{N^{.5}} + b_2 \overline{N} \tag{8.5}$$

where $\overline{N}$ is the mean of the treatment rates and $\overline{N^{.5}}$ is the mean of the square roots of the treatment rates,

$$\overline{N} = \frac{1}{n}\sum_{i=1}^{n} N_i \quad \text{and} \quad \overline{N^{.5}} = \frac{1}{n}\sum_{i=1}^{n} N_i^{.5} \tag{8.6}$$

and h_n is a constant calculated by

$$h_n = \frac{S_n^3}{S_n^2} + 2\overline{N^{.5}} \quad \text{where } S_n^2 = \sum_{i=1}^{n}\left(N_i^{.5} - \overline{N^{.5}}\right)^2$$

$$\text{and } S_n^3 = \sum_{i=1}^{n}\left(N_i^{.5} - \overline{N^{.5}}\right)^3 \tag{8.7}$$

For the special case when values for $N_i^{.5}$ are evenly spaced $h_n = 2\overline{N^{.5}}$ because then $S_n^3 = 0$.

This calculation procedure for obtaining yield variable values from a quadratic regression is summarized in Table 8.5.

Table 8.5 Calculation of the yield variables $\overline{Y}$, A_n and B_n from the single nutrient quadratic regression $Y = b_0 + b_1N^{.5} + b_2N$.

$Y = b_0 + b_1N^{.5} + b_2N$
$B_n = b_2$
$A_n = b_1 + b_2h_n$
$\overline{Y} = b_0 + b_1\overline{N^{.5}} + b_2\overline{N}$

An accurate procedure for computing the constants for these equations is illustrated in Table 8.6 with the example treatment rates P = 0, 15, 30, 60, giving the values $\overline{P} = 26.25$, $\overline{P^{.5}} = 4.274$ and $h_p = 7.4661$. Using the above equations with these constants and the coefficients of the quadratic regressions in Table 8.2 gives the identical values to those given by the regressions in Tables 8.3 and 8.4. For example with the regression $Y = 102.9 + 304.8P^{.5} + 4.013P$ for experiment A,

$$B_p = b_2 = 4.103$$

$$A_p = a_1 = b_1 + b_2 h_p = 304.8 + (4.013)(7.4661) = 334.8$$

$$\overline{Y} = b_0 + b_1 \overline{P^{.5}} + b_2 \overline{P}$$

$$= 102.9 + (304.8)(4.274) + (4.013)(26.25) = 1511.0$$

Table 8.6 Calculation of $\overline{P}$, $\overline{P^{.5}}$ and h_p for the treatment rates P_i.

	P_i	$P_i^{.5}$	$\left(P_i^{.5} - \overline{P^{.5}}\right)$	$\left(P_i^{.5} - \overline{P^{.5}}\right)^2$	$\left(P_i^{.5} - \overline{P^{.5}}\right)^3$
	0	0.0	-4.27404	18.26745	-78.07589
	15	3.87298	-0.40106	0.16085	-0.06451
	30	5.47723	1.20318	1.44765	1.74178
	60	7.74597	3.47192	12.05425	41.85142
Sum	105	17.09618	0.00000	31.93019	-34.54720
Mean	26.25	4.274			

$\overline{P} = 26.25$, $\overline{P^{.5}} = 4.274$ and $h_p = \dfrac{-34.54720}{31.93019} + 2(4.274) = 7.4661$

Inconsistent treatment rates

The need for consistent treatments for the computation of values for the yield variables is obvious in the case of $\overline{Y}$, estimated by the mean of the data from each experiment, because the mean of experimental data must vary with different sets of treatments except in the special case where the treatments have no effect. Similarly estimates of regression coefficients and hence of the other yield variables will vary with sets of treatment rates except in the special case that the regression model gives a perfect representation of the relationship that is being estimated.

Sometimes, for some practical reason, treatment rates vary amongst a series of experiments that would otherwise be suitable for the establishment of a general soil fertility model. The problem caused by the inconsistent rates can be overcome however, with the above recommended procedure, if it can be assumed that the higher degree regressions, from which the yield variable values are derived, approximate sufficiently closely to the actual fertilizer-yield relationships for the estimated regressions not to be significantly affected by the variations in the treatment rates. For example if an experiment has been carried out with the treatment rates 0, 18, 28, 64 for the nutrient N rather than a nominated standard set of rates 0, 15, 30, 60, it seems reasonable to assume that estimates of coefficients for the quadratic yield function would be essentially the same as those which would have been obtained if the nominated rates had actually been used. On this basis yield variable data can be calculated to obtain estimates of the values that would have been obtained if the standard rates had in fact been used. Thus for this example, the value for b_2 obtained from regressions $Y = b_0 + b_1N^{.5} + b_2N$ with the non-standard rates would be used as data values for B_n and data values for $\overline{Y}$ and A_n would be calculated from the regression, all as estimates of the values that would have been obtained if the standard rates had actually been used. It is important to note however that the assumption may not be justified when treatment rates vary markedly from a standard set.

Multinutrient regressions

The above procedure for calculating coefficients for regressions of lower degree from those for a quadratic regression are readily extended to multinutrient regressions as shown in Tables 8.7 to 8.9, the only new feature being the calculation of the mean of the square root of cross products

$$\overline{N^{.5}P^{.5}} = \frac{1}{nm}\sum_{i=1}^{n}\sum_{j=1}^{n} N_i^{.5}P_j^{.5} \tag{8.8}$$

8.3 Data features represented by yield variables

The yield variables $\overline{Y}$, A_p and B_p correspond directly to orthogonal components of a detailed analysis of variance and consequently they are important because they can be identified with meaningful and independent features of the yield data of fertilizer experiments, or in more general terms, to distinctive aspects of soil fertility as measured by the experiments. The nature of these features or aspects of soil fertility are illustrated with graphs in Figs 8.2 to 8.5 for the range of values in Table 8.10 obtained from the example data.

$\overline{Y}$ for level of yield

The variable $\overline{Y}$ is simply the mean of the yield data for each experiment and, as such, serves as an overall measure of the level of yield produced by the experiments as indicated by the graphs for the equations $Y = \overline{Y}$ in Fig. 8.2. As previously noted (6.21), $\overline{Y}$ relates directly to the zero trend coefficient in orthogonal polynomial "regressions" $p_0 Z = \overline{Y}$ differing only by a constant factor so that $\overline{Y} = g_{00} p_0$ where $g_{00} = Z = 1/n^{.5}$.

Table 8.7 Calculation of the yield variables $\overline{Y}$, A_n, A_p, B_{np}, B_n and B_p from a two nutrient quadratic regression.

$$Y = b_0 + b_1 N^{.5} + b_2 P^{.5} + b_3 (NP)^{.5} + b_4 N + b_5 P$$

$$B_p = b_5,\ B_n = b_4,\ B_{np} = b_3$$

$$A_p = b_2 + b_3 \overline{N^{.5}} + b_5 h_p$$

$$A_n = b_1 + b_3 \overline{P^{.5}} + b_4 h_n$$

$$\overline{Y} = b_0 + b_1 \overline{N^{.5}} + b_2 \overline{P^{.5}} + b_3 \overline{N^{.5} P^{.5}} + b_4 \overline{N} + b_5 \overline{P}$$

Table 8.8 Calculation of yield variables from a three variable quadratic regression.

$$Y = b_0 + b_1 N^{.5} + b_2 P^{.5} + b_3 K^{.5} + b_4 (NP)^{.5} + b_5 (NK)^{.5} + b_6 (PK)^{.5} + b_7 N + b_8 P + b_9 K$$

$$B_k = b_9,\ B_p = b_8,\ B_n = b_7,\ B_{pk} = b_6,\ B_{nk} = b_5,\ B_{np} = b_4$$

$$A_k = b_3 + b_5 \overline{N^{.5}} + b_6 \overline{P^{.5}} + b_9 h_k$$

$$A_p = b_2 + b_4 \overline{N^{.5}} + b_6 \overline{K^{.5}} + b_8 h_p$$

$$A_n = b_1 + b_4 \overline{P^{.5}} + b_5 \overline{K^{.5}} + b_7 h_n$$

$$\overline{Y} = b_0 + b_1 \overline{N^{.5}} + b_2 \overline{P^{.5}} + b_3 \overline{K^{.5}} + b_4 \overline{N^{.5} P^{.5}} + b_5 \overline{N^{.5} K^{.5}} + b_6 \overline{P^{.5} K^{.5}} + b_7 \overline{N} + b_8 \overline{P} + b_9 \overline{K}$$

Table 8.9 Calculation of yield variables from a four nutrient quadratic.

$$Y = b_0 + b_1N^{.5} + b_2P^{.5} + b_3K^{.5} + b_4S^{.5} + b_5(NP)^{.5} + b_6(NK)^{.5} + b_7(NS)^{.5}$$
$$+ b_8(PK)^{.5} + b_9(PS)^{.5} + b_{10}(KS)^{.5} + b_{11}N + b_{12}P + b_{13}K + b_{14}S$$

$$B_s = b_{14} \ , \ B_k = b_{13} \ , \ B_p = b_{12} \ , \ B_n = b_{11}$$

$$B_{ks} = b_{10} \ , \ B_{ps} = b_9 \ , \ B_{pk} = b_8 \ , \ B_{ns} = b_7 \ , \ B_{nk} = b_6 \ , \ B_{np} = b_5$$

$$A_s = b_4 + b_7\overline{N^{.5}} + b_9\overline{P^{.5}} + b_{10}\overline{K^{.5}} + b_{14}h_s$$

$$A_k = b_3 + b_6\overline{N^{.5}} + b_8\overline{P^{.5}} + b_{10}\overline{S^{.5}} + b_{13}h_k$$

$$A_p = b_2 + b_5\overline{N^{.5}} + b_8\overline{K^{.5}} + b_9\overline{S^{.5}} + b_{12}h_p$$

$$A_n = b_1 + b_5\overline{P^{.5}} + b_6\overline{K^{.5}} + b_7\overline{S^{.5}} + b_{11}h_n$$

$$\overline{Y} = b_0 + b_1\overline{N^{.5}} + b_2\overline{P^{.5}} + b_3\overline{K^{.5}} + b_4\overline{S^{.5}} + b_5\overline{N^{.5}P^{.5}} + b_6\overline{N^{.5}K^{.5}} + b_7\overline{N^{.5}S^{.5}}$$
$$+ b_8\overline{P^{.5}K^{.5}} + b_9\overline{P^{.5}S^{.5}} + b_{10}\overline{K^{.5}S^{.5}} + b_{11}\overline{N} + b_{12}\overline{P} + b_{13}\overline{K} + b_{14}\overline{S}$$

Table 8.10 Data values for yield variables obtained from the fertilizer experiment data of Table 8.1 and the site variable T for soil phosphorus.

Experiment	Yield variables			Site variable
	$\overline{Y}$	A_p	B_p	T (ppm P)
A	1511.0	334.8	4.013	1
B	2100.0	244.2	-11.25	2
C	2200.0	164.7	31.94	3
D	2382.5	92.37	-25.85	5
E	2802.5	57.97	-8.648	12
F	2985.0	2.885	-1.656	24
G	3295.0	-28.69	-8.663	78

A_p for yield response to fertilizer

The yield variable A_p is the value for the coefficient a_1 in the regression $Y = a_0 + a_1 P^{.5}$ as estimated by the regressions in Table 8.3 and as such provides an overall measure of the yield response to the fertilizer treatments. It relates directly to the linear trend coefficient p_1 in orthogonal polynomial regressions $Y = p_0 Z + p_1 L$, differing only by a constant factor, $A_p = a_1 = g_{11} p_1$ (by equation 6.22). The nature of the trend is seen by subtracting graphs for the zero trend, $Y = \overline{Y}$ in Fig. 8.2 from graphs for the linear regressions (square root scale) of Table 8.2 in Fig. 8.3 to obtain the linear trends in Fig. 8.4. The responses thus illustrated range from a large positive response for experiment A to a small negative response for experiment G, corresponding to the range of values for A_p or a_1, 334.8 to –28.69.

B_p for curvature in fertilizer-yield regressions

The yield variable B_p is the value for the coefficient b_2 in the regression $Y = b_0 + b_1 P^{.5} + b_2 P$ as estimated by the regressions in Table 8.2 and provides a measure of the curvature in this regression additional to that provided by the linear trend regressions for $Y = a_0 + a_1 P^{.5}$ as illustrated in Fig. 8.5 with graphs obtained by subtracting the graphs for the linear regression in Fig. 8.3 from the corresponding graphs for the quadratic regressions in Fig. 8.1. It relates directly to the quadratic trend coefficient p_2 in the orthogonal regressions $Y = p_0 Z + p_1 L + p_2 Q$ differing only by a constant factor, $B_p = b_2 = g_{22} p_2$ (by equation 6.23). Values of B_p or b_2 for this example range from negative to positive but are mostly very small (the vertical scale in Fig. 8.5 is ten times that of the other figures) corresponding in this respect to the non-significant quadratic effects in the analyses of variance. They can consequently be regarded as mainly representing error effects in the data for each experiment.

B_{np} for interaction

For multinutrient models, yield variables corresponding to those illustrated above with the example are computed for each nutrient in turn and in addition yield variables for interaction effects. An interaction is illustrated in Fig. 6.2 by the three dimensional graphs for the $p_3 L_n L_p$ trend. Again the yield variables relate directly to the orthogonal trend coefficient in corresponding orthogonal polynomial regressions, differing only by a constant.

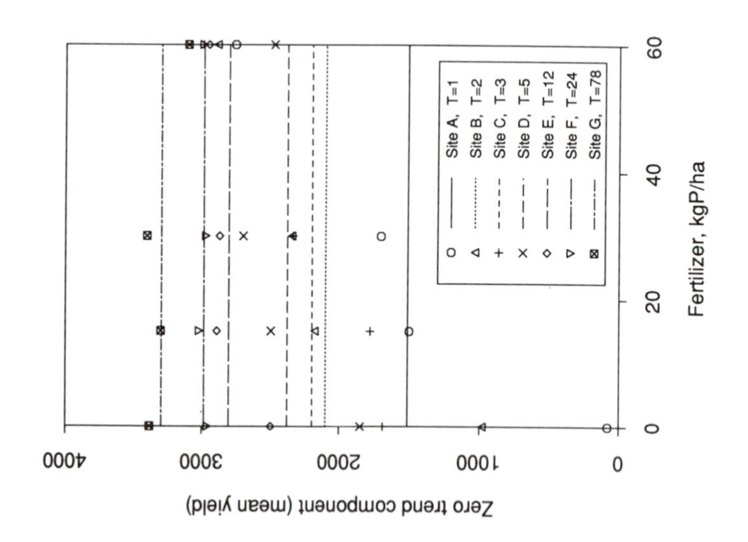

Fig. 8.2 Graphs for estimates of the "regressions" $Y = \overline{Y}$ used to obtain data values for the yield variable $\overline{Y}$.

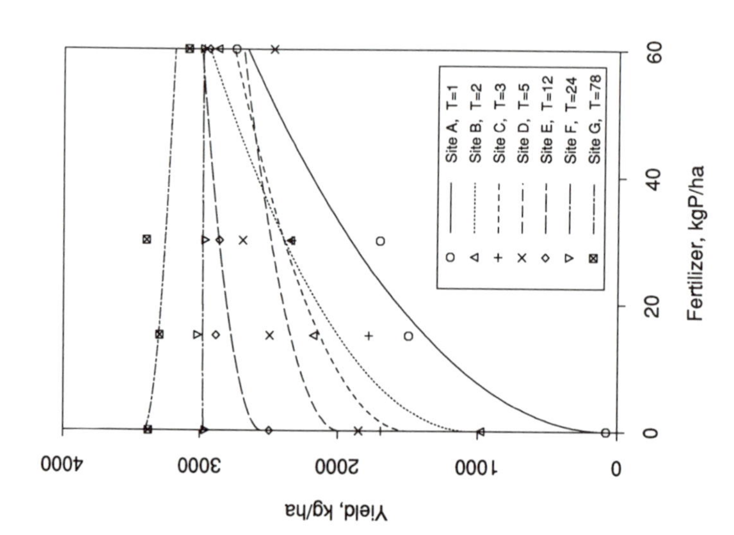

Fig. 8.3 Graphs for estimates of the regressions $Y = a_0 + a_1N^{.5}$ used to obtain data values for the yield variable A_p (= a_1).

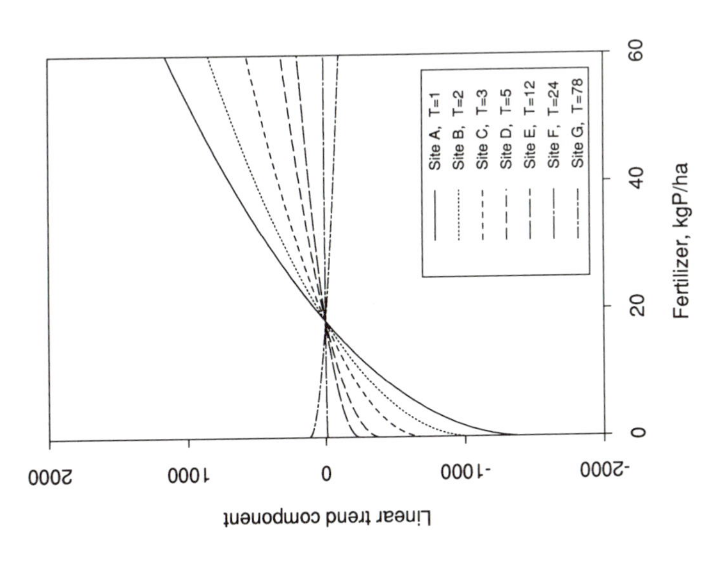

Fig. 8.4 Graphs for the linear trend in the polynomial regressions of Table 8.2.

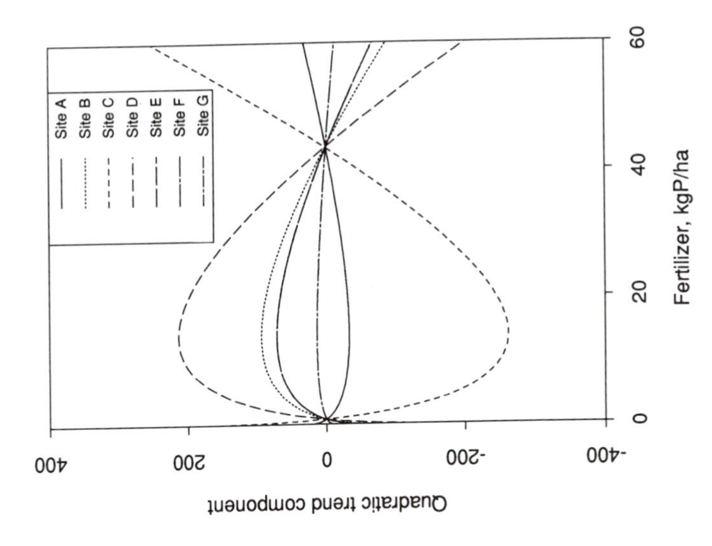

Fig. 8.5 Graphs for the quadratic trend in the polynomial regressions of Table 8.2.

8.4 Development of a general model

Relationships with site variables

The development of general models to represent relationships between yield and site variables involves regression studies. In the case of the introductory example these are simple because there is only one site variable, T, for the soil test for the level of soil phosphorus at the experiment sites before the establishment of the experiments, so that all that is necessary is to select suitable regression forms to represent the relationships with this variable and then to estimate the regressions. Normally there will be many site variables because many variables can affect the results of experiments in addition to soil nutrient level so that studies are required to select appropriate variables and to choose appropriate forms for the regressions, as described in the following chapter. For the example a simple study of the yield variable and site variable data in Table 8.10 leads to the regressions

$$\overline{Y} = 1207.3 + 585.91T^{.5} - 39.901T \qquad (8.9)$$

for the relationship between yield level ($\overline{Y}$) and the site variable T for soil nutrient level and

$$A_p = 48.302 + 310.69T^{-1} - 1.554T \qquad (8.10)$$

for the relationship between yield response to the fertilizer (A_p) and T, the form of the regression in each case being chosen on the basis of previous experience with similar relationships. These are suitable for a general model because they are statistically significant ($p<0.001$) with respective R^2 values 94.8% and 96.3%, and R_a^2 values 92.2% and 94.5%, and because they represent the types of relationship that are expected from knowledge about the nature of the variables. Thus the graph for the $\overline{Y}$ regression (8.9) in Fig. 8.6 shows effects of soil nutrient level T on level of yield corresponding to the common form of fertilizer-yield functions and the graph for the A_p regression (8.10) in Fig. 8.7 shows the expected type of decrease in yield response to P fertilizer with increase in soil nutrient level.

Corresponding studies on the relationship between B_p for curvature and T do not indicate any significant relationship suggesting either that some other site variable is responsible for the variations in curvature or that any relationship is too small relative to the error variance to be apparent. In the absence of a statistically significant relationship, B_p values are best estimated by the mean of the data values giving the estimating equation $B_p = -2.8763$.

The set of equations thus derived from the series of fertilizer experiments,

$$\left.\begin{aligned} \overline{Y} &= 1207.3 + 585.91T^{.5} - 39.901T \\ A_p &= 48.302 + 310.69T^{-1} - 1.1554T \\ B_p &= -2.8763 \end{aligned}\right\} \tag{8.11}$$

constitute a general soil fertility model. They represent the relationships between variations in aspects of soil fertility that have been quantified by the yield variables and the only available measure of the causes of this variability in this case, that is the site variable T for soil phosphorus. The adequacy of the model in these respects depends on the extent to which the experiments represent the variability of the soil fertility in the region and the extent to which this variability is explained by the regressions of the yield variables on the site variables. With this example, the high R_a^2 values for the $\overline{Y}$ and A_p regressions indicate that the models are highly successful in accounting for the variability in yield level and yield response to the P fertilizer and the absence of a significant relationship for B_p, a failure to explain the small variations in curvature. It is as well to remember however that the example is based on a set of data for only seven experiments that have been selected to introduce the procedure for developing a model rather than to provide a model that represents a region of any importance. Models for extensive and variable regions require data from many more experiments and with many more site variables as described in the following chapter.

8.5 Predictions with a general model

Prediction of fertilizer-yield functions

The set of equations relating yield variables to site variables that constitute a general model are developed from studies of the relationships between the results of fertilizer experiments, as represented by the yield variables, and site variables and they consequently may be used to predict corresponding results for particular sites by substituting values for the site variables. Since the set of yield variables correspond to orthogonal components of fertilizer-yield functions, such estimates constitute collectively an estimate of the fertilizer-yield functions that would be estimated for particular sites if

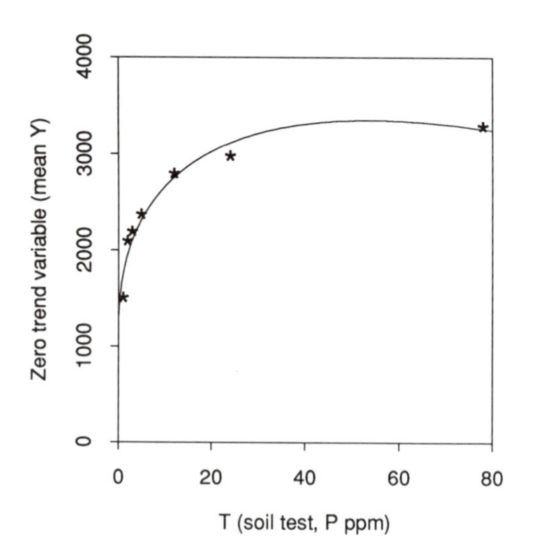

Fig. 8.6 Graph for $\overline{Y} = 1207.3 + 585.91T^{.5} - 39.901T$, regression (8.9), with data *.

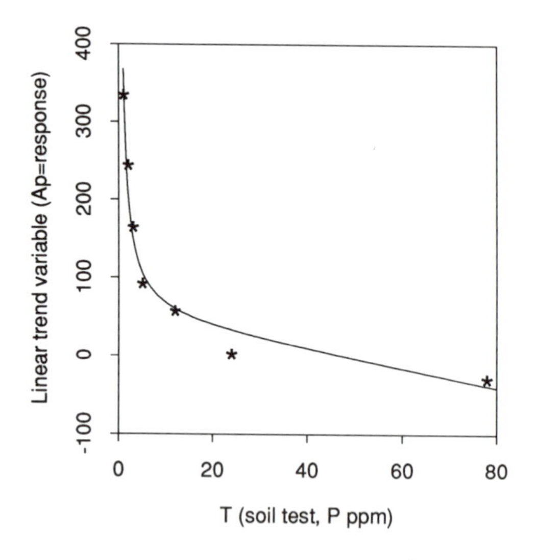

Fig.8.7 Graph for $A_p = 48.302 + 310.69T^{-1} - 1.554T$, regression (8.10) with data *.

fertilizer experiments were actually carried out at those sites. Equations that convert estimates of the yield variables into estimates of fertilizer-yield functions are given in Tables 8.11 to 8.13 for one, two and three nutrient functions, being derived from the equations in Tables 8.5, 8.7 and 8.8 for calculating the yield variable values from yield functions. The "hats" on the yield variables in these equations, for Y , A_n , etc., are added to indicate that the yield variable values are statistical estimates provided by the general model equations. The equations can be readily extended for four or more nutrient functions, if general models were ever developed for such multinutrient relationships.

The use of the Table 8.11 equations can be illustrated with the example general model (8.11) and the values $h_p = 7.4661$, $\overline{P^{.5}} = 4.274$ and $\overline{P} = 26.25$ that were originally derived for the calculation of the yield variables (Table 8.6). For example if a yield function is to be

Table 8.11 Calculation of a single nutrient yield function from general model estimates of yield variables.

$\hat{Y} = b_0 + b_1 N^{.5} + b_2 N$ where
$b_2 = \hat{B}_n$
$b_1 = \hat{A}_n - b_2 h_n$
$b_0 = \hat{\overline{Y}} - b_1 \overline{N^{.5}} - b_2 \overline{N}$

Table 8.12 Calculation of a two nutrient yield function from general model estimates of yield variables.

$Y = b_0 + b_1 N^{.5} + b_2 P^{.5} + b_3 (NP)^{.5} + b_4 N + b_5 P$ where
$b_5 = \hat{B}_p$, $b_4 = \hat{B}_n$, $b_3 = \hat{B}_{np}$
$b_2 = \hat{A}_p - b_3 \overline{N^{.5}} - b_5 h_p$
$b_1 = \hat{A}_n - b_3 \overline{P^{.5}} - b_4 h_n$
$b_0 = \hat{\overline{Y}} - b_1 \overline{N^{.5}} - b_2 \overline{P^{.5}} - b_3 \overline{NP^{.5}} - b_4 \overline{N} - b_5 \overline{P}$

Table 8.13 Calculation of a three nutrient yield function from general model estimates of yield variables.

$$Y = b_0 + b_1N^{.5} + b_2P^{.5} + b_3K^{.5} + b_4(NP)^{.5} + b_5(NK)^{.5} + b_6(PK)^{.5}$$
$$+ b_7N + b_8P + b_9K$$

where

$$b_9 = \hat{B}_k,\ b_8 = \hat{B}_p,\ b_7 = \hat{B}_n,\ b_6 = \hat{B}_{pk},\ b_5 = \hat{B}_{nk},\ b_4 = \hat{B}_{np}$$

$$b_3 = \hat{A}_k - b_5\overline{N^{.5}} - b_6\overline{P^{.5}} - b_9h_k$$

$$b_2 = \hat{A}_p - b_4\overline{N^{.5}} - b_6\overline{K^{.5}} - b_7h_p$$

$$b_1 = \hat{A}_n - b_4\overline{P^{.5}} - b_5\overline{K^{.5}} - b_7h_n$$

$$b_0 = \hat{\overline{Y}} - b_1\overline{N^{.5}} - b_2\overline{P^{.5}} - b_3\overline{K^{.5}} - b_4\overline{NP^{.5}} - b_5\overline{NK^{.5}} - b_6\overline{PK^{.5}}$$
$$- b_7\overline{N} - b_8\overline{P} - b_9\overline{K}$$

estimated for T = 3, the values $\hat{B}_p$ = −2.8736, $\hat{A}_p$ = 148.40 and $\hat{\overline{Y}}$ = 2102.4 are predicted by the general model equations and substituting these values in the Table 8.11 equations gives the estimated yield function

$$\hat{Y} = 1451.9 + 169.9P^{.5} - 2.874P$$

for sites with T = 3. Corresponding estimates can be obtained for other values of the site variable as illustrated below in Table 8.14.

Variable parameter form of general models

An extension of the above procedure for calculating estimates of yield functions from values for site variables is obtained by substituting the general model regression equations for the yield variables in the conversion equations (Table 8.11 to 8.13) and then gathering and rearranging the terms to obtain a *variable parameter form of the general model.* For example given general model equations $b_0 = f_0(T)$, $b_1 = f_1(T)$ and $b_2 = f_2(T)$ derived for the site variable T from a series of yield functions $Y = b_0 + b_1N^{.5} + b_2N$, the function can be generalized by replacing the coefficients b_0 , b_1 , b_2 by the functions to obtain the variable parameter form of the model

$$\hat{Y} = f_0(T) + f_1(T)N^{.5} + f_2(T)N$$

Terms in this equation can then be collected and rearranged in a convenient form for calculating direct estimates of the yield function for nominated values for the site variable T. Thus if the example general model equations (8.11) are represented for convenience by

$$\overline{Y} = u_0 + u_1T^{.5} + u_2T$$

$$A_p = v_0 + v_1T^{-1} + v_3T$$

$$B_p = w_0$$

then substituting the predictive expressions for $\overline{Y}$, A_p and B_p in the relationships of Table 8.11 gives equations for the fertilizer-yield function coefficients

$$b_2 = w_0$$

$$b_1 = v_0 + v_1T^{-1} + v_2T - b_2h_p$$

$$b_0 = u_0 + u_1T^{.5} + u_2T - b_1\overline{P^{.5}} - b_2\overline{P}$$

corresponding to the above $b_2 = f_2(T)$, $b_1 = f_1(T)$ and $b_0 = f_0(T)$. Substituting for b_1 and b_2 in the right hand expressions

$$b_2 = w_0$$

$$b_1 = v_0 + v_1T^{-1} + v_2T - w_0h_p$$

$$b_0 = u_0 + u_1T^{.5} + u_2T - (v_0 + v_1T^{-1} + v_2T - w_0h_p)\overline{P^{.5}} - w_0\overline{P}$$

and rearranging gives the variable parameter generalization of $Y = b_0 + b_1P^{.5} + b_2P$ in the form $Y = C_0 + C_1P^{.5} + C_2P$ where

$$C_0 = u_0 - (v_0 - w_0h_p)\overline{P^{.5}} - w_0\overline{P} + u_1T^{.5} - v_1\overline{P^{.5}}T^{-1} + (u_2 - v_2\overline{P^{.5}})T$$

$$C_1 = v_0 - w_0h_p + v_1T^{-1} + v_2T$$

$$C_2 = w_0$$

Substituting coefficient values from the example general model equations (8.11) and the values for h_p, $\overline{P^{.5}}$, and $\overline{P}$ from the Table 8.6 thus gives the variable parameter form of the example general model

$$\hat{Y} = (984.6 + 585.9T^{.5} - 1328T^{-1} - 34.96T) + (69.76 + 310.7T^{-1} - 1.155T)P^{.5} - 2.874P \quad (8.12)$$

Variable parameter forms of general models are convenient when there are only a few variables involved as in this example but they become very large and cumbersome when the models are for multinutrient yield functions or when there are many site variables.

Smoothing of error effects

General models are based on regressions that have been estimated from the data of many experiments and because of this, estimates of yield functions obtained from the models benefit from a smoothing of the error effects in the yield functions for the individual experiments. Estimates of yield functions obtained from general models are consequently usually preferable to estimates obtained from the data of individual experiments. Such a smoothing of error effects is illustrated by comparing the functions in Table 8.2 and Fig. 8.1 that have been estimated for individual experiments with those in Table 8.14 and Fig. 8.8 that have been estimated from the model, with corresponding values for the site variable *T*. The general model estimates show a much more orderly and consistent sequence for the range of the site variable than the corresponding estimates for individual experiments. In particular the function estimated from the general model estimate for

Table 8.14 General model estimates of the yield variables $\bar{Y}$, A_p, B_p and coefficients for the function $Y = b_0 + b_1P^{.5} + b_2P$ calculated from these estimates. Values for the site variable *T* correspond to those for the experimental data in Table 8.1.

T	$\hat{\bar{Y}}$	$\hat{A}_p$	$\hat{B}_p$	b_0	b_1	b_2
1	1753	357.8	−2.874	208	379.3	−2.874
2	1956	201.3	−2.874	10792	222.8	−2.874
3	2102	148.4	−2.874	1452	169.9	−2.874
5	2318	104.7	−2.874	1854	126.1	−2.874
12	2758	60.3	−2.874	2484	81.8	−2.874
24	3120	33.5	−2.874	2961	55.0	−2.874
78	3269	−37.8	−2.874	3415	−16.4	−2.874

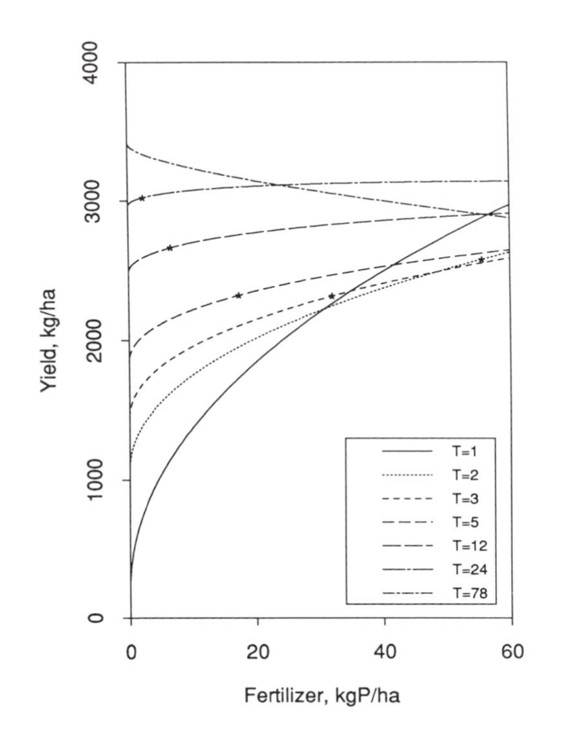

Fig. 8.8 Graphs for the fertilizer-yield functions in Table 8.2, estimated individually from the data for each experiment in Table 8.1, * = optimal fertilizer rate.

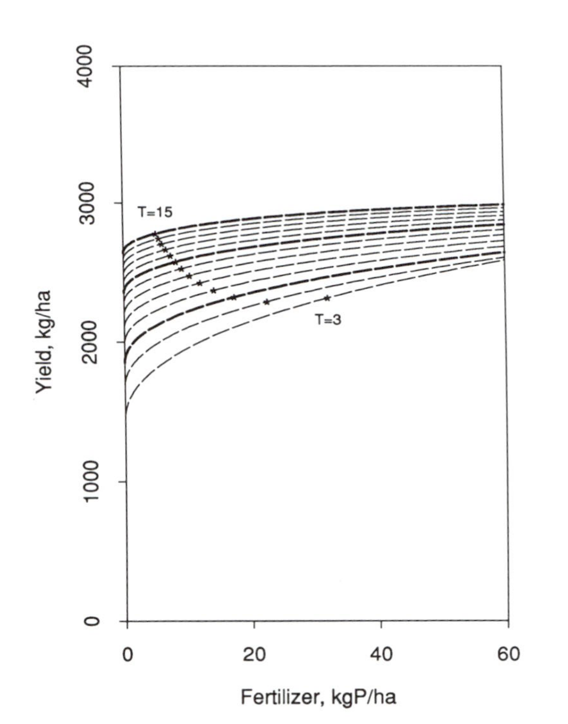

Fig. 8.9 Graphs for the fertilizer-yield functions with soil test T = 3 to 15, estimated from the general model (8.11) or (8.12), * = optimal fertilizer rate.

$T = 3$ shows the usual diminishing response form and not the abnormal form apparently caused by an error effect for experiment C where also $T = 3$.

Prediction of optimal fertilizer application rates

Yield functions estimated from general models can be used in the same ways as functions estimated directly from the data of individual experiments to calculate yields, profits and optimal fertilizer application rates. Again values calculated from functions estimated from general models show more orderly sequences than corresponding values calculated directly from yield functions for individual experiments due to a smoothing of error effects. This is shown for example by the optimal rates indicated by the asterisks on the graphs of estimated functions in Fig. 8.9 for the range $T = 3$ to 15, calculated by $dY / dP = E_p$ with $E_p = 12$ for the economic variables, avoiding extrapolated estimates for $T < 3$. For farmer advisory services, a series of optimal rates can be calculated in this way for a convenient range of values for the site variable to obtain calibrations for soil tests in terms of optimal fertilizer rates using E_p values chosen to suit local economic conditions, as illustrated in Fig. 8.9 using the example value $E_p = 12$.

8.6 Invalid procedures

The important feature of the procedure for deriving general models is that it is based on yield variables that represent *orthogonal* components of the variance of the data for the individual experiments. Thus for the example, model equations with the forms

$$\overline{Y} = u_0 + u_1 T^{\cdot 5} + u_2 T$$

$$A_p = v_0 + v_1 T^{-1} + v_2 T$$

$$B_p = w_0$$

were established by regression studies with the yield variables $\overline{Y}$, A_p, B_p and the site variable T. The fact that such equations can be combined into a single equation in the form of a variable parameter function

$$Y = (c_0 + c_1 T^{\cdot 5} + c_2 T^{-1} + c_3 T) + (c_4 + c_5 T^{-1} + c_6 T) P^{\cdot 5} + c_7 P \qquad (8.13)$$

as detailed above for the example equation (8.12) can mislead by suggesting two seemingly obvious and simpler alternative procedures:

1. To estimate general models such as (8.13) directly from the data of a series of fertilizer experiments by the ordinary multiple regression procedure.
2. To estimate the varying parameter expressions for a general model by estimating regression relationships between the coefficients of ordinary yield functions and site variables.

There are important statistical reasons for not using either of these procedures:

1. *Objections to direct estimation with multiple regressions*

Regressions of yield variables on site variables can be combined to derive a general model in the form of a variable parameter function such as the above example (8.13) where the function $Y = b_0 + b_1P^{.5} + b_2P$ is generalized by replacing the parameter b_0 by $(c_0 + c_1T^{.5} + c_2T^{-1} + c_3T)$ and b_1 by $(c_4 + c_5T^{-1} + c_6T)$. Such generalized forms of yield functions can be estimated, validly and directly, by fitting a multiple regression to data for crop yield, fertilizer application rates and site variables obtained from a series of fertilizer experiments. The fact that when this is done the identical model is obtained to that obtained by the yield variable procedure, if corresponding sets of variables are used as regressor variables for the alternative procedures, is important in that it demonstrates that the yield variable procedure provides a least squares estimate of the relationships. However, there is a serious statistical problem with this direct estimation procedure. Standard tests of significance on multiple regressions estimated from this type of data are invalid because the data contains several very different types of error. Consequently the residual mean sum of squares from the standard regression analysis of variance is not an estimate of an error variance but is a rather meaningless value obtained from a combination of very different error effects, each with its own error variance.

Firstly there is a *within* experiment error variance associated with the data for each experiment, giving a series of different within experiment error variances, and then there is a very different *between* experiment error variance corresponding to the variations in the results of the individual experiments that cannot be accounted for by the site variables. As explained in chapter 4, *within* experiment error is produced by factors that vary within the experiment site, or within blocks for blocked designs, such as the soil, effects of weeds or pests, and the *between* experiment error is produced by these factors plus other factors such as soil composition, kind of soil, the weather, agricultural practices that are essentially constant within individual experiments but differ, often greatly, amongst different experiments. Although site variables can be included in regressions to allow for the effects of such factors, the procedures for measuring the factors are mostly imperfect so that there remain substantial unexplained so-called error effects. Within site error variances are usually very much smaller than the between site error variance so that the combination of all of these error deviations in the residual mean square of an analysis of variance of a regression produces an invalid estimate of error variance, much smaller than that which should be used to test the significance of effects from site variables. This is particularly unfortunate because it results in an over estimation of the significance of the effects of site variables so that researchers, excited by apparently highly significant results, are reluctant to accept the fact that the tests are invalid and that their results are not nearly as wonderful as they thought they were. Consequent problems are compounded if a weighted regression procedure is used, weighting data by the inverse of the within site error variances.

This problem of heterogeneity in estimates of error variance is avoided by the use of the yield variable procedure because the within site error variances associated with yield variable values are usually of negligible magnitude relative to the between site error variance for regressions of these variables on site variables. Consequently standard tests of significance for the regressions of yield variables on site variables based on the residual mean squares of the regression analyses of variance are valid.

Another problem with the use of the direct multiple regression procedure, more scientific than statistical, is that appropriate general model forms for the effects of the site variables on crop yield cannot be readily chosen on the basis of scientific knowledge about the nature of the effects. The model form (8.13) for the example was indicated by the studies of the relationship of the yield variables to site variables and would not otherwise have been apparent. Although variable parameter models may be derived by trial selections of variables, it is not easy to make selections based on scientific knowledge about the nature of the variables and their effects.

2. *Objections to regressions for parameters of non-orthogonal functions*

The variable parameter form of general models may also suggest to the unwary that the expressions that replace the coefficients can be estimated directly by regressions of the yield function coefficients on site variables. Thus for the example, the variable parameter form of the model, $Y = (c_0 + c_1T^{.5} + c_2T^{-1} + c_3T) + (c_4 + c_5T^{-1} + c_6T)P^{.5} + c_7P$, as a generalization of the function $Y = b_0 + b_1P^{.5} + b_2P$ may suggest that estimates of the bracketed expressions may be obtained simply and directly with the regressions

$$b_0 = c_0 + c_1T^{.5} + c_2T^{-1} + c_3T$$

$$b_1 = c_4 + c_5T^{-1} + c_6T$$

where data values for the coefficients b_0 and b_1 are obtained by regressions estimated from the data of a series of experiments as listed in Table 8.2. Such a procedure parallels that followed with yield variables but with the important difference that regressions for yield functions such as $Y = b_0 + b_1P^{.5} + b_2P$ are not orthogonal. Consequently the values of the coefficients b_0 and b_1 do not correspond directly to orthogonal components of the variance of the yield data for each experiment. Unless the dependent variables for regressions on site variables correspond to orthogonal components of the variance of the data for each experiment, the combination of the regressions in the form of general models will give invalid and misleading estimates of the relationships between the results of fertilizer experiments and site variables in variable regions. Scientists may find this requirement of orthogonality particularly difficult to accept when they can perceive perfectly sensible relationships between the values of coefficients for non-orthogonal regressions and site variables. In particular it may be difficult to understand that the coefficients of the Mitscherlich model in the popular form

$$Y = a\left[1 - \exp\{c(P + b)\}\right]$$

cannot be used to obtain general models by developing regressions for the model parameters on site variables because they do not also correspond to orthogonal components of the yield data from which they have been estimated, even though biological or physical meanings can be associated with each of the parameters and sensible regression relationships can be established for each parameter with site variables.

The fact that regressions of non-orthogonal yield function coefficients on site variables cannot be combined to obtain estimates of general models is well documented (for example Maddala 1977) and is easily demonstrated by comparing estimates of the general relationships obtained in this way with direct multiple regression estimates of the same relationship. Thus with the example, using the values for b_0 and b_1 given by the regressions in Table 8.2 as data for dependent variables and the site variable values of Table 8.1, regression estimates of the relationships

$$b_0 = c_0 + c_1T^{.5} + c_2T^{-1} + c_3T \text{ and } b_1 = c_4 + c_5T^{-1} + c_6T$$

are

$$b_0 = 1290.9 + 495.2T^{\cdot 5} - 166.7T^{-1} - 29.13T \quad \text{and}$$

$$b_1 = 90.42 + 228.8T^{-1} - 0.8904T$$

and these give very different estimates of the bracketed expressions in the general model

$$Y = (c_0 + c_1T^{\cdot 5} + c_2T^{-1} + c_3T) + (c_4 + c_5T^{-1} + c_6T)P^{\cdot 5} + c_7P$$

to those derived either from the regressions for the yield variables $\overline{Y}$ and A_p or the direct estimate by a multiple regression, that is to

$$b_0 = 984.6 + 585.9T^{\cdot 5} - 1328T^{-1} - 34.96T \quad \text{and}$$

$$b_1 = 69.76 + 310.7T^{-1} - 1.155T$$

in equation (8.12).

Corresponding problems can be demonstrated for general models for multinutrient yield functions to be described in chapter 9, although less readily because the derivation of varying parameter forms of the general model from the yield variable regressions involves much larger algebraic expressions.

The basis for the identity between general models derived by the combination of regressions of yield variables on site variables and corresponding direct regression estimates, as described above and with mathematical detail in Appendix B, follows from:

1. The direct algebraic relationship between corresponding orthogonal and non-orthogonal polynomial regressions described in chapter 6, for example between regressions for the yield function models $Y = b_0 + b_1P^{\cdot 5} + b_2P$ and $Y = p_0Z + p_1L + p_2Q$.
2. The fact that the coefficients of orthogonal regressions can be estimated independently of each other.

The yield variables differ from the coefficients of the orthogonal polynomial regressions only by a constant, $\overline{Y} = g_{00}p_0$ by (6.21), $A_p = g_{11}p_1$ by (6.22) and $B_p = g_{22}p_2$, so that correspondingly (i) estimates of the yield variables can be used to calculate estimates of the non-orthogonal regression coefficients and (ii) estimates of the yield coefficients can be obtained independently of each other. The yield variable procedure is based on these features of orthogonal polynomial regressions and the relationships with non-orthogonal regressions.

Identity with non-orthogonal models

The identity between general model coefficients derived by the combination of regressions of trend coefficients on site variables and those derived by a single multiple regression of yield on site variables depends essentially on two conditions:

1. The use of the orthogonal regressions (8.14) for the fertilizer-yield relationship.
2. The choice of corresponding sets of site variable for the alternative derivations.

This last condition, if not satisfied, can cause confusion. Thus for the above example the varying parameter form of the general model

$$Y = (c_0 + c_1 T^{\cdot 5} + c_2 T^{-1} + c_3 T) + (c_4 + c_5 T^{-1} + c_6 T) P^{\cdot 5} + c_7 P \qquad (8.14)$$

was derived from the yield variable equations

$$\overline{Y} = u_0 + u_1 T^{\cdot 5} + u_2 T$$

$$A_p = v_0 + v_1 T^{-1} + v_2 T$$

$$B_p = w_0 \qquad (8.15)$$

and the same form, but with different coefficient values is derived by the same procedure from the alternative set of yield variable equations

$$\overline{Y} = u_0 + u_1 T^{\cdot 5} + u_2 T + u_3 T^{-1}$$

$$A_p = v_0 + v_1 T^{-1} + v_2 T$$

$$B_p = w_0 \qquad (8.16)$$

where the regressor variable T^{-1} has been added to the $\overline{Y}$ regression. In the example the estimate of (8.19), using the regression $\overline{Y}$ = 1207.3 + 585.91T$^{\cdot 5}$ − 39.901T in the general model (8.11) was

$$Y = (984.6 + 585.9T^{\cdot 5} - 1328T^{-1} - 34.96T) + (69.76 + 310.7T^{-1} - 1.155T)P^{\cdot 5} - 2.874P \qquad (8.17)$$

whereas if the regression

$$\overline{Y} = 2088.1 + 260.8T^{\cdot 5} - 13.92T - 805.4T^{-1} \qquad (8.18)$$

is used the estimate becomes

$$Y = (1865.4 + 260.8T^{.5} - 2133.3T^{-1} - 8.986T) + (69.76 + 310.7T^{-1} - 1.155T)P^{.5} - 2.874P \qquad (8.19)$$

This second equation is identical with the direct regression estimate of (8.14) demonstrating that the regressions (8.16) must be used to obtain the identity demonstrated above with the orthogonal regressions. In fact both models give very similar estimates of yield but the first estimate (8.17) is preferable because the $\overline{Y}$ regression without the additional variable T^{-1} has a higher R_a^2 value due to a lower residual mean square in the analysis of variance.

These complications, although mathematically interesting, are rather academic and only arise if comparisons are made of single equation versions of general models that have been estimated by alternative procedures. The above comparisons with orthogonal regressions are sufficient to demonstrate that the general model derived by regressions of the yield variables on site variables is identical with that obtained with the direct least squares regression procedure. Corresponding identities do not exist for regressions of non-orthogonal regression coefficients on site variables and they should consequently not be used to derive general models.

Chapter 9

Development and Use of General Soil Fertility Models

Summary

General soil fertility models are basically sets of regression equations that relate aspects of the results of fertilizer experiments to site variables. Their distinctive feature is that the "independent aspects of the results" of the experiments are represented by yield variables that correspond to orthogonal components of analyses of variance of regressions for fertilizer-yield relationships estimated from the data of the experiments in order that regressions for the yield variable-site variable relationships can be used jointly to estimate corresponding fertilizer experiment results in the form of fertilizer-yield functions. The estimated functions can then be used to calculate yields and profits for nominated fertilizer application rates and fertilizer requirements from values for the site variables in exactly the same way as if the functions had been obtained by carrying out fertilizer experiments.

To develop a general model:

1. The soil fertility of a region is sampled by carrying out a series of fertilizer experiments with a design suitable for the estimation of fertilizer-yield functions.
2. The results of each experiment are represented by regressions for fertilizer-yield functions of appropriate form that are estimated from the experimental data.
3. Yield variables that both represent meaningful features of the experimental results and correspond to orthogonal components of analyses of variance of the fertilizer-yield functions, are computed.
4. Data are obtained for site variables at each experiment site that are expected to explain the variations in the yield variable values for the different experiments.

5. Regression equations are developed that relate the yield variables to site variables, giving the set of equations that constitute the general model. In the absence of significant regressions for yield variables, the "regression" equations are simply the equations with the form $V = constant$ where *constant* is the mean of the yield variable V.

To use a general model:

1. Values for the site variables of the model are measured or estimated for particular sites and substituted in the model equations to obtain estimates of the yield variables.
2. The estimated yield variables are used to calculate estimates of fertilizer-yield functions of the same form as those estimated from the original experimental data.
3. The estimated yield functions are used to calculate estimates of yields, profits, optimal fertilizer rates, etc. using the same procedures as those used with regressions estimated directly from experimental data.

Although these procedures for the development and use of models are essentially those for the development and use of regressions as described in many excellent references (for example Gunst and Mason 1980, Weisberg 1980, Draper and Smith 1981, Cook and Weisberg 1982, Atkinson 1985), there are traps for the unwary. The simple use of standard regression procedures without an appreciation of their mathematical bases and limitations, and of the scientific nature of the relationships being estimated, can easily produce misleading and inaccurate estimates.

9.1 Yield variable data

Data for a general soil fertility model are obtained by carrying out a series of experiments at sites that are selected to represent the range of soil fertility that occurs in a region, providing in effect a sampling of the soil fertility of the region. All of the experiments must have the same design and treatments, to provide a consistent set of data. This uniformity of treatments is important since, for example, varying treatments to match expected variations in fertilizer requirements, as suggested by soil analyses, will produce a misleading bias towards the experimenter's expected results - small variations in treatment rates can be tolerated however by estimating yield variable values for a standard set of rates as described in section 8.2. Thus designs from section 7.6, selected for the number of nutrient deficiencies to be represented by the model, with an appropriate set of fertilizer treatment rates, can be used to obtain sets of experimental data for yield variables based on the following fertilizer-yield functions, as summarized in Table 9.1.

$$Y = b_0 + b_1N^{.5} + b_2N + b_3L_b + b_4Q_b + b_5C_b \qquad (9.1)$$

$$Y = b_0 + b_1N^{.5} + b_2P^{.5} + b_3(NP)^{.5} + b_4N + b_5P + b_6L_b \qquad (9.2)$$

$$Y = b_0 + b_1N^{.5} + b_2P^{.5} + b_3K^{.5} + b_4(NP)^{.5} + b_5(NK)^{.5} + b_6(PK)^{.5}$$
$$+ b_7N + b_8P + b_9K + b_{10}L_b + b_{11}Q_b \qquad (9.3)$$

$$Y = b_0 + b_1N^{.5} + b_2P^{.5} + b_3K^{.5} + b_4(NP)^{.5} + b_5(NK)^{.5} + b_6(PK)^{.5}$$
$$+ b_7N + b_8P + b_9K + b_{10}L_b \qquad (9.4)$$

$$Y = b_0 + b_1N^{.5} + b_2P^{.5} + b_3K^{.5} + b_4S^{.5} + b_5(NP)^{.5} + b_6(NK)^{.5}$$
$$+ b_7(NS) + b_8(PK)^{.5} + b_9(PS)^{.5} + b_{10}(KS)^{.5} + b_{11}N + b_{12}P$$
$$+ b_{13}K + b_{14}S + b_{15}L_b + b_{16}Q_b + b_{17}C_b \qquad (9.5)$$

Although ideally all important aspects of fertility would be represented by the yield variables, the fact that experiments increase in size with the number of nutrients used for treatments and that site

Table 9.1 Yield variables for yield functions (9.1) to (9.5) that have been estimated from experiments with designs of chapter 7. Calculation formulas are in Tables 8.5 to 8.9 of chapter 8.

Yield function	Chap. 7 Design	Chap. 8 Table	Yield variables
(9.1)	3	8.5	$\bar{Y}$, A_n, B_n
(9.2)	4	8.7	$\bar{Y}$, A_n, A_p, B_{np}, B_n, B_p
(9.3)	5	8.8	$\bar{Y}$, A_n, A_p, A_k, B_{np}, B_{nk}, B_{pk}, B_n, B_p, B_k
(9.4)	6	8.8	$\bar{Y}$, A_n, A_p, A_k, B_{np}, B_{nk}, B_{pk}, B_n, B_p, B_k
(9.5)	7	8.9	$\bar{Y}$, A_n, A_p, A_k, A_s, B_{np}, B_{nk}, B_{ns}, B_{pk}, B_{ps}, B_{ks}, B_n, B_p, B_k, B_s

variable data must be obtained to relate to each of the yield variables, imposes practical limitations. Consequently general model projects are best restricted to experiments with a maximum of two or three nutrients and hence to a maximum of 6 or 10 yield variables for aspects of soil fertility. If there are more than two or three soil nutrient deficiencies, the less important of these may be removed by applying uniform applications of appropriate fertilizers to the experiment sites before establishing the experiments and then developing general models for variations in soil fertility, in the absence of these deficiencies. The consequent restriction in the scope of the model will not be important for the estimation of fertilizer requirements if the excluded deficiencies are not sufficiently severe to produce interaction effects that affect the yield variables other than $\overline{Y}$ for level of yield, except of course in so far as the models will not give information about the effects of the excluded deficiencies on yield levels.

9.2 Site variable data

Many site variables may affect the yield variable values that are obtained with fertilizer experiments and ideally experiments will be located to represent the range of all of these, and all of their combinations, that occur in a region. Usually however it will not be practicable to obtain data at all sites for all of the variables that may conceivably be responsible for the variations in the magnitudes of the yield variables and consequently selections of variables must be made on the bases of their expected importance and available research resources.

Sampling

The major problem in obtaining data for site variables is undoubtedly that of sampling the experiment sites so that the data values will adequately represent the sites. Poorly representative samples can produce poor relationships that are often attributed, quite unjustly, to inadequacies in the methods of analysis. Thus soil analyses for sites often vary widely both with location laterally across the sites, and with depth down the soil profile in the site, so that it is difficult to obtain single analysis values that will represent the site. With relatively immobile nutrients such as phosphorus, soil analyses are often much higher in the surface cultivation horizon due to the accumulation of residues from previous fertilizer applications or from plant litter so that analyses of samples of this horizon can suffice, the effects of the nutrients at greater depths being relatively small. Moreover, since mixtures of samples usually give analysis values that are close to the means of analyses of the individual samples in the mixture, a well mixed sample of many sub samples from the surface horizon can be expected to give analyses close to the mean analysis for the site, even when there are large variations due to the effects of residues from previous fertilizer applications. However for other nutrients that exist in a highly mobile form, such as the nitrate anion form of soil nitrogen, profile distributions may vary greatly so that allowance must be made for the vertical distribution of analyses within the root zone as well as for lateral variations across sites. Although there are statistical procedures for representing trends of analyses down soil profiles in the form of site variables, as described below, the need to also allow for lateral variations, by collecting and mixing samples from many profiles, presents considerable practical difficulties. Some site variables may also vary with time during the growth of a crop as for example analyses for nitrogen, sulphur or manganese, due to the effects of factors such as microbial activity, leaching by rainfall and nutrient uptake by the growing crop, so that analyses made at any particular time may not be satisfactory for site variables. In such cases it may be better to use more general site variables such as the amounts of previous fertilizer applications or dummy variables (described below) for types of recent cropping.

Apart from variables associated with the soil, the weather that happened to prevail at sites during the course of experiments can have profound effects on their results. Accordingly site variables that represent weather effects must also be devised as best possible. Recording meteorological data for such variables over a period of time may be impracticable however, because of the expense involved in the purchase and maintenance of recording equipment, so that indirect provision for weather effects may be necessary as described below.

Surface soil analyses

The effects of fertilizers on crop production vary with the amounts of soil nutrients that are available to plants and there are many simple methods of analysis for soil nutrients, called soil tests, that when carried out on surface soil samples, serve as guides to fertilizer requirements. Such analyses or tests usually constitute the more important of the site variables for general soil fertility models. Since however the analyses *relate to* rather than actually *measure* the amounts of soil nutrients that are available for plant growth, they have to be interpreted with allowance for the factors that affect their relationship to nutrient uptake by plants such as the chemical and physical properties of the soils, weather conditions and soil microbial activity. This is done by extending the general model regressions to include additional site variables for such other factors. Thus for example, simple regressions for the relationship of the yield variable A_n for yield response to N fertilizer with the soil nutrient analysis or test T for soil nitrogen as measured by water extractable nitrate

$$A_n = f(T)$$

may be improved in some situations by the extension of the function to include variables R and pH

$$A_n = f(T, R, pH)$$

where R is a site variable for rainfall received during some period of the plant growth and pH is a site variable for soil pH. The choice of appropriate functional forms and variables for such regressions requires knowledge both about the nature of the variables used as site variables and the factors that affect their relationships with yield variables, and this may require supplementary research studies. Research may also be required to determine the most appropriate procedure for measuring site variables. For example chemically very different procedures for soil phosphorus analysis may give near equivalent regressions with yield variables if appropriate soil analyses for other soil properties are included in the regressions (Colwell and Donnelly 1971) so that different selections of site variables are possible. Examples of regressions with several site variables are given below.

Soil profile trends

Soil properties and the relative magnitude of their effects on plant growth vary with their depth down soil profiles. For example with increase in depth down the soil profile and within the root zone of crops, salt content may increase from harmless to toxic concentrations, or plant available potassium may increase from deficiency to sufficiency levels. In such cases site variables have to be devised that will both represent the distribution of the feature within soil profiles and allow for the variations in the importance of its effects with depth. Since soils are commonly classified and characterized on the basis of features of their morphological horizons, it may seem only logical, at least to pedologists, to do this by using horizon analyses as site variables. Such analyses are usually unsatisfactory for general models however because the thickness, depth or even the occurrence of

Table 9.2 Soil profile analyses for water extractable nitrate (ppm N). The variable D for regressions on depth is the mean for each sampling depth.

Sampling depth (cm)	D (cm)	Site 1 (ppm N)	Site 2 (ppm N)	Site 3 (ppm N)	Site 4 (ppm N)
0-10	5	16.3	35.1	44.4	8.0
10-20	15	21.4	27.0	17.5	10.6
20-30	25	19.9	11.6	13.5	11.6
30-40	35	11.6	7.9	9.4	9.1
40-50	45	11.3	7.6	6.2	9.6
50-60	55	13.8	7.0	3.9	10.4
60-70	65	16.0	6.2	1.8	9.9
70-80	75	18.3	4.6	0.7	9.9
80-90	85	18.8	3.6	0.7	11.8
90-100	95	19.2	2.9	1.7	11.9

morphological horizons varies amongst sites so that it is not possible to obtain consistent sets of data. Because of the need for consistency, soil profiles must be sampled consistently, using a standard set of depths such as 0 to 5, 5 to 10, 10 to 20, 20 to 30, 30 to 50 cm irrespective of the obvious horizons of the soil profile. Morphological horizons, although closely associated with some important physical features of soils, are in any case often not closely associated with chemical features as can be demonstrated by detailed analyses of profiles at successive depths, covering the transitions between horizons as well as trends within horizons.

Soil analyses for a succession of depths down a soil profile are not suitable by themselves for site variables because the successive analyses are usually correlated with each other. Moreover the form of the trends in analyses with depth vary, as shown by the example profile analyses in Table 9.2 for the water extractable nitrate anion form of soil nitrogen at experiment sites (Colwell 1977) and such variations may affect relationships with yield variables. For these reasons, site variables are required for profile trends. These can be obtained by representing the profile analyses by orthogonal polynomial regressions[1] for depth

[1] For example regressions for the characterization and comparison of soil profiles, see Colwell (1970).

$$A = p_0\xi_0 + p_1\xi_1 + p_2\xi_2 + p_3\xi_3 + \cdots \qquad (9.6)$$

where A is soil analysis and the ξ_0, ξ_1, ξ_2, ξ_3, ... are orthogonal polynomial expressions for zero, linear, quadratic, cubic, ... trends with depth. Such a representation corresponds to those described in chapters 6 and 7 for representing trends in yield with fertilizing rate by orthogonal polynomial regressions. Thus for the example profile data in Table 9.2 the coefficients for the respective regressions

$$A = 52.68\xi_0 + 0.8588\xi_1 + 6.267\xi_2 + 0.5160\xi_3$$

$$A = 35.89\xi_0 - 27.21\xi_1 + 15.00\xi_2 - 7.772\xi_3$$

$$A = 31.56\xi_0 - 32.53\xi_1 + 19.57\xi_2 - 9.144\xi_3$$

$$A = 32.51\xi_0 + 2.103\xi_1 + 0.4961\xi_2 + 1.858\xi_3$$

can be used as data values for variables representing zero, linear, quadratic and cubic trends in the respective analysis profiles of experiment sites. More conveniently these same trends can be represented by the underlined coefficients in the corresponding series of non-orthogonal polynomial regressions

$$A = \underline{a_0}\bar{A}$$

$$A = b_0 + \underline{b_1}D$$

$$A = c_0 + c_1D + \underline{c_2}D^2$$

$$A = d_0 + d_1D + d_2D^2 + \underline{d_3}D^3 \qquad (9.7)$$

where $\bar{A}$ is mean analysis, the underlined coefficients corresponding to orthogonal trend coefficients as with yield variables, differing only by constants[2]. The values for regression coefficients obtained in this way by non-orthogonal polynomial regressions for the example profiles in Table 9.3 consequently show corresponding variations in relative magnitudes to those shown by the trend coefficients in the orthogonal polynomial regressions. Thus the coefficients of orthogonal regressions of form (9.6) relate to the coefficients of the regressions (9.7), $a_0 = 03162p_0$, $b_1 = -0.01101p_1$, $c_2 = 0.0004352p_2$, $d_3 = 0.00001799p_3$ and may be used as data values for site variables to represent independent features of the profile trends.

---

[2] Values for the constants are determined by the relative spacing of the values of the polynomial variable as described in chapter 6 for the relationships of the form $\bar{Y} = g_{00}p_0$, $a_1 = g_{11}p_1$, $b_2 = g_{22}p_2$, $c_{33} = g_{33}p_3$, equations (6.21) to (6.24).

Table 9.3 Underlined coefficients for regressions (9.7) corresponding to the orthogonal trend coefficients of (9.6) . These may be used as site variables to represent independent aspects of the trends in the profile analyses in Table 9.2.

Trend	Coefficient	Site 1	Site 2	Site 3	Site 4
Zero	a_0	16.66	11.35	9.980	10.28
Linear	b_1	0.009454	−0.2996	−0.3582	0.02315
Quad.	c_2	0.002727	0.006527	0.008515	0.0002159
Cubic	d_3	0.00000929	−0.0001398	−0.0001645	0.00003343

Site variables for profile analyses may be used for sequential studies on regression relationships with yield variables, commencing with regressions with the site variable for the zero trend with depth (the profile mean analysis) and followed by regressions with the addition of variables of progressively higher degree, until improvements to the regression become non-significant. Corresponding studies may be made with regressions on other scales, for example the square root scale, and with regressions for profiles to other depths, to determine the scale and depth most appropriate for this type of site variable.

Regressions of yield variable on orthogonal coefficients p_0 , p_1 , ... of (9.6) for trends with depth, corresponding to the coefficients a_0 , b_1 , ... in (9.7), can be used to calculate *weighting vectors* that indicate the relative importance of analyses at different depths down the soil profile and the nature of the relationships. The derivation of the vectors is shown conveniently using these orthogonal trend coefficients and matrix notation. Thus if a regression of the form

$$\overline{Y} = b_0 + b_1 p_0 + b_2 p_1 + b_3 p_2 + b_4 p_3 \tag{9.8}$$

for mean yield $\overline{Y}$ on the site variables p_0 , ... , p_3 is estimated, then the expression $b_1 p_0 + b_2 p_1 + b_3 p_2 + b_4 p_3$ can be represented as the vector product

$$\mathbf{b'p} \tag{9.9}$$

Since the vector $\mathbf{p}$ for the coefficients in the orthogonal regression (9.6) is derived by $\mathbf{p} = \mathbf{E'a}$, where $\mathbf{E}$ is the matrix of orthogonal polynomial trends and $\mathbf{a}$ is the vector of analyses, this expression may be written

$$\mathbf{b'p} = \mathbf{b'E'a}$$
$$= \mathbf{w'a} \qquad (9.10)$$

where $\mathbf{w} = \mathbf{Eb}$ is the weighting vector. The vector product $\mathbf{w'a}$ gives in effect a weighted sum of the profile analyses for the best regression relationship with the yield variable $\overline{Y}$ so that the magnitudes of the elements in $\mathbf{w}$ are a direct statistical estimate of the relative importance of the analyses down the soil profile. Note however that the calculation requires the use of the orthogonal matrix $\mathbf{E}$ and the trend coefficient vector $\mathbf{p}$ so that the weighting vector is not readily calculated from the corresponding non-orthogonal values for $\overline{Y}$, a_1, b_2, ... However, significant relationships with profile trend variables are often restricted to only zero (mean) and linear trends so that the weighting vectors merely indicate the rather obvious fact that analyses near the soil surface have relatively greater effects on plant growth than those at greater depths (Colwell 1979).

Dummy variables for classifications

Experiment sites can be classified on the basis of descriptive and non-quantifiable features such as kind of soil, location in an agricultural district, location in a climatic zone, or type of agricultural practices, and the classifications can be represented by variables in regressions with the device of *dummy variables* whereby a variable is defined as having the values 1 or 0 according to whether the site belongs or not to a particular classification unit. For example if experiments had been carried out in a region containing the five kinds of soil S1, S2, S3, S4 and S5, dummy variables for the soil classification would be defined as in Table 9.4.

Table 9.4 Example dummy variable data values for the five soil classifications S1 to S5.

Kind of	Value of dummy variable				
soil	*S1*	*S2*	*S3*	*S4*	*S5*
S1	1	0	0	0	0
S2	0	1	0	0	0
S3	0	0	1	0	0
S4	0	0	0	1	0
S5	0	0	0	0	1

Dummy variables can be used as variables in regressions with $n - 1$ variables for n classifications, each with a value of either 0 or 1. For example if there are five kinds of soil, the relationship between the yield variable $\overline{Y}$ and soil can be represented by the regression

$$\overline{Y} = b_0 + b_1 S2 + b_2 S3 + b_3 S4 + b_4 S5 \quad (9.11)$$

where *S2, S3, S4* and *S5* are four of the five dummy variables defined in Table 9.4. The use of $n - 1$ variables for n classifications is necessary because one of the classifications is defined if the value for all of the other $n - 1$ variables is known. Thus for the example if $S2 = S3 = S4 = S5 = 0$, then the soil must be *S1* with the value $\overline{Y} = b_0$. The excluded dummy variable, *S1* for the example, serves as a reference since the coefficients for the other dummy variables indicate their difference from this reference value. Thus soil *S1* serves as a reference with the estimated value b_0 for $\overline{Y}$ and corresponding estimates for the other soils are respectively $\overline{Y} = b_0 + b_1$, $\overline{Y} = b_0 + b_2$, $\overline{Y} = b_0 + b_3$ and $\overline{Y} = b_0 + b_4$, differing from this reference value by the value of the respective coefficients, b_1 for the difference between $\overline{Y}$ values for soil S1 and S2, b_2 for the difference between soil S1 and S3, and so on.

Weather

The weather is obviously important as a site variable but is also difficult to represent by simple site variables because both it and its effects on crop growth vary with time. For example a single frost during a critical stage of growth can have devastating effects on crop production and rainfall at certain times has much more important effects on growth than at others. Moreover the regular measurement of even the simplest of weather variables such as daily rainfall and temperature may become prohibitively expensive if there are many experiments as required for the development of general models. Weather effects are so important however that even very simple measurements can be valuable as site variables. Thus dummy variables for location within climatic districts or the coefficients of orthogonal polynomial trends with time for rainfall received in regular intervals during the growing period can be useful, as described by Fisher (1924) and used for example by Colwell and Morton (1984).

The problems caused by the impracticality of obtaining site variable data for the weather for use in general models, although important, should not be exaggerated. Useful models can be developed in the absence of any weather variables because weather effects are greater for the yield variable $\overline{Y}$ for level of yield than the other yield variables for fertilizer treatment effects. Consequently models without site variables for weather effects can be used to represent the effects of site variables on yield variables for fertilizer effects and, since level of yield does not affect the calculation of optimal rates, to estimate fertilizer requirements. Also if data have been obtained over several years and at many sites so that weather effects approximate to those of a random variable, regressions without weather variables will provide estimates of relationships for average growing conditions. In any case farmers must grow their crops on the basis of expected weather conditions so that they require estimates for average conditions.

9.3 Example general models

Two examples are given to illustrate the development of general models from data provided by series of fertilizer experiments and the way they may be used to estimate fertilizer requirements in the region represented by the experiments. The first is derived from a very simple project that was planned to provide a calibration for estimating the phosphorus fertilizer requirements of wheat from soil test analyses for phosphorus (Colwell and Esdaile 1968) and the second is from a much more ambitious project that was planned to establish bases for estimating nitrogen and phosphorus fertilizer requirements of wheat from soil analyses, expected rainfall and soil classification in a large region comprising four distinct agricultural districts (Colwell 1977, 1979; Colwell and Morton 1984). Experience gained with these and similar projects led to the development of the procedures described in this book.

Example 1. A simple soil test calibration

Stage 1: Data collection

The original objective of the project for the first example was to develop a calibration for estimating phosphorus fertilizer requirements of wheat from soil test analyses for phosphorus, for a wheat growing region of the state of New South Wales, Australia. To this end a series of fertilizer experiments was planned with wheat with the phosphorus fertilizer treatment rates 0, 4.9, 9.8, 19.6 kg P/ha,[3] to be established at

[3] Treatment rates were expressed in imperial units as 0, 50, 100, 200 lb/acre.

sites in the region covering a range of soil phosphorus levels as indicated by surface soil analyses by a soil testing procedure. Then, in addition, it was decided to investigate the need for nitrogen fertilizer in the region by applying a nitrogenous fertilizer at the rates 0, 51.6, 103.2 kg N/ha in factorial combination with the P treatments and to also obtain a surface soil analysis for pH since this soil property might affect the availability of fertilizer phosphorus to plants and hence the calibration of the P test analyses. Eventually then, yield and soil analysis data were obtained from 47 experiments over a period of three successive years at sites distributed throughout the region, all with the same treatments in a 3 × 4 factorial design with two replicates in blocks. The experiments thus provided a consistent set of data suitable for the development of a general model.

Data for the yield variables $\overline{Y}$ for level of yield, A_n and A_p for yield responses to the N and P fertilizer treatments, B_{np} for N × P interaction effect and B_n and B_p for quadratic curvature in the N and P responses are obtained from the experiment data for each of the 47 sites by first estimating regressions of the form

$$Y = b_0 + b_1 N^{\cdot 5} + b_2 P^{\cdot 5} + b_3 (NP)^{\cdot 5} + b_4 N + b_5 P + b_6 L_b$$

and then using the equations in Table 8.7 with the constants $N^{\cdot 5} = 5.7807$, $N = 51.6$, $P^{\cdot 5} = 2.4428$, $P = 8.5750$, $N^{\cdot 5} P^{\cdot 5} = 14.121$, $h_n = 9.6092$ and $h_p = 4.2672$ as determined by the treatment rates. No use is made of the coefficient b_6 for block effect since the objective is to estimate mean functions for whole sites that can be related to site variables, also for whole sites. Since the variable L_b has the values − 1 and + 1 for location in the blocks of each site (Table 7.2), mean site regressions are obtained by substituting the mean $L_b = 0$, that is by simply deleting the $b_6 L_b$ from the regressions.

Data for the site variables T and pH were obtained directly by the phosphorus soil test and pH analyses for each site. Since the data values were to represent the whole of the experiment sites, site analyses were obtained using a well mixed soil sample composed of many sub samples that have been collected from each site. Note incidentally that there is no need for analyses of samples from the individual plots of experiments and if such data were obtained, at great cost if there are many plots, they would simply be averaged to obtain mean analyses for the sites. For practical purposes the mean of analyses of individual soil samples from a site is the same as the analysis of a mixture of the same samples, even for pH despite the fact that this analysis represents the negative logarithm of hydrogen ion activity.

Stage 2: Regression studies

Regression studies for this example, on the relationships between the yield variables and site variables, are simple because there are only two site variables.

$\overline{Y}$ *regressions* Knowledge about the effects of soil phosphorus and pH on plant growth indicates that either or both of the site variables, T for soil phosphorus or pH for soil pH, may be expected to relate to the levels of yield for the experiments, represented by the yield variable $\overline{Y}$, but there is no established theory to indicate the mathematical form of the relationships apart from the expectation that the relationship between yield and soil nutrient level will be similar to that for fertilizer-yield functions. Nor was there any local experience to indicate which of these variables would have the more important relationship with $\overline{Y}$ in the region of the experiments. Accordingly the data were used to estimate a series of regressions with various alternative forms such as

$$\overline{Y} = b_0 + b_1 T \tag{9.12}$$

$$\overline{Y} = b_0 + b_1 T^{.5} \tag{9.13}$$

$$\overline{Y} = b_0 + b_1 T^{.5} + b_2 T \tag{9.14}$$

$$\overline{Y} = b_0 + b_1 pH \tag{9.15}$$

$$\overline{Y} = b_0 + b_1 T + b_2 pH \tag{9.16}$$

$$\overline{Y} = b_0 + b_1 T^{.5} + b_2 pH \tag{9.17}$$

$$\overline{Y} = b_0 + b_1 T^{.5} + b_2 T + b_3 pH \tag{9.18}$$

with the objective of finding the regression showing the most meaningful relationship as indicated by R_a^2 values and tests of significance. For this example, these studies only served to produce the single significant relationship

$$\overline{Y} = 0.2835 + 0.3007\,pH\,, \quad R_a^2 = 6.5^*\% \tag{9.19}$$

corresponding to the form (9.15) and explaining 6.5% of the variance of $\overline{Y} = b_0 + b_1 T$. Although significant (*, p<0.05), the low R_a^2 value indicates that the regression only explains a very small proportion of the variance of $\overline{Y}$ so that it is only useful as a very general guide to yield level, there obviously being other important sources of variation not covered by the investigation. Fortunately level of yield does not affect estimates of optimal fertilizer rates so that the poor relationship did not affect the success of the project in achieving its original objective of providing a basis for estimating P fertilizer requirements with the soil test T.

A_p regressions The A_p variable represents yield response to P fertilizer so that its magnitude is expected to decrease with increase in soil P as represented by the soil test variable T. Also since the solubility of soil phosphorus is known to increase with increase in soil pH, A_p values may be expected to decrease with increase in values for the variable pH. Studies on regression relationships, similar to those described above for $\overline{Y}$, produced the highly significant ($p<0.001$) regression relationship

$$A_p = 0.7245 - 0.1058T^{.5} + 0.005737T - 0.03945pH, \quad R_a^2 = 48.2^{***}\% \qquad (9.20)$$

in accordance with these expectations based on existing scientific knowledge about the nature of the variables. Note however that the regression only accounts for 48.2% of the variation of A_p so that there are still other important sources of variation not included in the regression.

B_p regressions Corresponding regression studies with the variable B_p for curvature in the response to P fertilizer applications are completely empirical since there is no scientific information to indicate variables that can be expected to affect curvature, apart from the expectation that they would be the same as those affecting yield response, nor to indicate the form of relationships that should be estimated. The best regression estimated in this case, as indicated by tests of significance and comparisons of R_a^2 values, was

$$B_p = 0.03144 + 0.0003835T - 0.007325pH, \quad R_a^2 = 13.5^{*}\% \qquad (9.21)$$

Although the quadratic curvature as represented by this yield variable affects calculations of optimal fertilizer rates from square root fertilizer-yield functions, the low R_a^2 and level of significance is not very important since analyses of variance show that in general the quadratic curvature represents only a small proportion of the yield variance, and a much smaller proportion than the linear component, represented by A_p.

A_n, B_{np} and B_n regressions Corresponding studies with the yield variable A_n, B_{np} and B_n produced no significant regressions and this may be explained by the fact that the N fertilizer treatments were

included in the project as an afterthought, simply to investigate the existence of nitrogen deficiencies in the region. For this reason no site variables were specifically included in the project that might relate to the effects of the N fertilizer applications. For this example the absence of significant relationships serves to illustrate the use of the yield variable means $\overline{A}_n$ = 0.02575, $\overline{B}_{np}$ = 0.003793 and $\overline{B}_n$ = − 0.0002493 as alternatives to regressions on site variables, in general models.

The regression studies thus produced the set of prediction equations, constituting a general soil fertility model, in Table 9.5.

Stage 3: Application

Substituting values for the site variables in the model regressions gives estimates of the yield variables $\overline{Y}$, A_p and B_p. These, together with mean values for the other yield variables and the equations in Table 8.12 can be used to calculate estimates of yield functions with the form

$$Y = b_0 + b_1 N^{.5} + b_2 P^{.5} + b_3 (NP)^{.5} + b_4 N + b_5 P$$

Table 9.5 A simple general soil fertility model consisting of regressions for the yield variables $\overline{Y}$, A_p and B_p on the site variables T_p for soil phosphorus (ppm P) and *pH* for soil pH, and sample means for A_n, B_{np} and B_n.

Trend	Prediction equation	R_a^2 (%)
Mean yield	$\overline{Y} = 0.2835 + 0.3007pH$	6.5*
Response to N	$A_n = 0.02575$	
Response to P	$A_p = 0.7245 - 0.1058T^{.5} + 0.005737T$ $- 0.03945pH$	48.2***
N×P interaction	$B_{np} = 0.003793$	
Curvature for N	$B_n = -0.0002493$	
Curvature for P	$B_p = 0.03144 + 0.0003835T$ $- 0.007325pH$	13.5*

(*, p<0.05; ***, p<0.001).

corresponding to that used with the regressions to represent the results of the original fertilizer experiments. In general terms, regression relationships between the results of fertilizer experiments and site variables can be used to predict corresponding results from values for the site variables. These estimates can then be used to calculate simultaneous optimal N and P fertilizer application rates (equations 3.10) using appropriate values for the economic variables. Thus even though the project was planned to simply give a soil test calibration for optimal application rates of P fertilizer with no attempt to measure any site variable that might relate to the effects of the N fertilizer treatments, the inclusion of the N treatments has had the consequence that both N and P rates can be estimated from the model because of inclusion of mean values for the variables A_n, B_{np} and B_n in the model. Example optimal rates obtained in this way with the example economic variable values E_n = 0.010 and E_p = 0.016 are given in Table 9.6.

Table 9.6 A soil test calibration for estimating optimal rates of application of fertilizer derived from the general model in Table 9.5.

pH	T_p (ppm P)			
	10	20	30	40
	Optimal kg N/ha Optimal kg P/ha			
5	4.9 48.7	3.6 28.2	2.8 15.9	1.8 5.5
6	3.2 22.2	2.3 10.3	1.7 4.0	1.1 0.5
7	2.5 12.2	1.8 4.9	1.3 1.5	0.9 0.1
8	2.0 7.5	1.5 2.7	1.1 0.6	0.9 0.0

The example calibration indicates a wide range of P fertilizer requirements, ranging from the high rate of 48.7 kg P/ha for T_p = 10 and pH = 5 to nil for T_p = 40 and pH = 8 but only a narrow range of N fertilizer requirements from the low rate of 4.9 kg N/ha to the very low rate of 0.9 kg N/ha for the same range of test values. Given the treatment rates of 0, 51.6, 103.2 kg N/ha and the features of the square root quadratic model (section 5.6) these estimates of optimal rates for N are not likely to be very accurate and since they are small, this may not seem to be important. Since however the model includes no site variable that relates to the yield variables A_n, B_{np} and B_n for the N fertilizer effects, the range of estimates of N rates is restricted, with rates varying because of the estimate of the mean linear N × linear P effect given by the mean of the variable B_{np}. The project thus provided the required P soil test calibration for estimating P fertilizer requirements but because of the way it had been planned, only a very general and inadequate indication of N fertilizer requirements.

Example 2. A general model with many site variables

This example is used both to illustrate the development of general models from extensive sets of data and to describe the types of problem that may be encountered with such a development.

Sampling problems

Ideally a general model is developed from yield variable data that have been obtained by an adequate sampling of the range of soil fertility in the region with fertilizer experiments and from site variable data that give an adequate measurement of all the factors that are responsible for the variations of these yield variable data. In practice however this ideal is only likely to be partially attained because of the limitations imposed by available research resources and existing scientific knowledge so that in practice models must be developed as best possible from admittedly inadequate data. The consolation in this respect is that the statistical procedures, if properly employed, will develop about the best models possible with the available data and scientific knowledge and hence provide statistical bases for about the best estimates of the yields and fertilizer requirements of crops in the region represented by the model. If alternative estimates prove to be better, such as those that may be based on what is often vaguely described as "local experience", then it follows that there are additional site variables that can be identified and used to further develop and improve the models.

Sampling inadequacies arise, rather inevitably, because of the following general features of fertilizer experiments:

1. Practical considerations limit the range of soil fertility and growing conditions that can be represented by fertilizer experiments and hence the adequacy of the yield variable data that can be obtained to represent a region. Experiments must be located at convenient and accessible sites and this often results in the selection of sites with the better farmers in the more developed locations of a region, producing a sampling bias towards the more developed and higher fertility parts of the region, and the bias is increased when experiments are rejected as failures because they have been adversely affected by unfavourable growing conditions such as drought, excessive rain, frost, poor management and so on.
2. Sampling with respect to time must be limited to at most a few years, if a project is to be completed within a reasonable time. Consequently the sampling of the seasonal growing conditions that occur in the region is likely to be inadequate. This problem may be overcome to some extent by basing estimates on meteorological records if these are extensive and if relationships can be established with the variables that have been recorded. In general however such records are not available or adequate for new or developing regions where the need for general models is greatest.
3. The site variables that can be used in models to explain the variations in the yield variables are inevitably restricted to those which can be measured with available resources and technology. Serious inadequacies in this respect often only become apparent after completion of the data acquisition phase of a project, when attempts are being made to interpret results.

Such problems should be expected and recognized in soil fertility projects for the development of general models but the consequent limitations on the usefulness of the models should not be exaggerated. Models covering the effects of many variables can be developed as a means for applying scientific knowledge as illustrated with the following example from a project on the nitrogen and phosphorus fertilizer requirements of wheat in the wheat growing districts of southern Australia (Hallsworth 1969; Colwell 1977, 1979; Colwell and Morton 1984).

Stage 1: Data collection

Sampling For the example, data have been selected for 139 fertilizer experiments that provide a sampling of the fertility of five very different kinds of soil, in five wheat growing districts, in an area

Table 9.7 Distribution of the 139 example experiments amongst five soils and five districts in a wheat growing region of southern Australia.

District	Soil				
	S1	S2	S3	S4	S5
D1	0	19	20	0	1
D2	10	0	3	4	0
D3	0	0	10	15	7
D4	0	0	17	0	0
D5	0	0	9	0	24

extending about 800 km through a wheat growing region of south eastern Australia. The sampling of the soils and districts with these experiments is indicated by the numbers of experiments for each combination of soil and district in Table 9.7, soils being denoted by S1, S2, ... and districts by D1, D2, ... for this example. A sampling problem is immediately apparent. Despite the large number of experiments most of the soils only occur in certain districts so that estimates for many of the soil-district combinations must be based onan inadequate amount of data or obtained by extrapolation from the data for other soil-district combinations.

Yield variables The experiments were planned to have the same design with the 16 factorial combinations of the treatment rates 0, 22.4, 56.2, 112.1 kg N/ha and 0, 11.2, 28.0, 56.0 kg P/ha, all replicated in three blocks so that the results of each could be represented by a regression with the form

$$Y = b_0 + b_1 N^{\cdot 5} + b_2 P^{\cdot 5} + b_3 (NP)^{\cdot 5} + b_4 N + b_5 P + b_6 L_b + b_7 Q_b$$

where L_b and Q_b are variables for linear and quadratic trends across the blocks with values −1, 0, +1 and +1, −2, +1 respectively for location in blocks (Table 7.2). Yield variable data values for $\overline{Y}$, A_n, A_p, B_{np}, B_n, B_p can be calculated from these regressions and treatment rates using the Table 8.7 equations and the constant values $\overline{N^{\cdot 5}} = 5.704$, $\overline{N} = 47.675$, $\overline{P^{\cdot 5}} = 4.030$, $\overline{P} = 23.80$, $\overline{N^{\cdot 5}P^{\cdot 5}} = 22.99$, $h_n = 10.346$ and $h_p = 7.313$. However most of the experiments had somewhat different treatment rates from the intended standard rates because of difficulties with the calibration of experimental machinery and some experiments had missing plots due to various events. Since however the treatment rates did not vary greatly from the intended rates and because there was an adequate replication, it was possible to assume that the same yield functions would have been estimated at each site if the intended rates had been applied and if there had been no missing data. Thus yield variable values corresponding to those that would have been obtained with the standard rates and complete data were calculated by estimating regressions from the available data and then calculating values using these constant values for the standard rates.

Site variables A wide range of site variable data were recorded and the following are selected for the present example:

- Soil analyses. Soil samples were taken from the surface soil and the soil profiles of each site prior to the establishment of the experiments and the samples were analysed for soil nutrients using a range of soil test analysis procedures, and for a range of chemical and physical properties, providing a wide range of site variable data. The surface cultivation soil layer of each site was sampled by taking about 200 random samples from the surface depth 0-10 cm and mixing to obtain a single composite sample for each of the sites. In addition nine soil profiles were sampled to a depth of 100 cm with the 10 cm sampling intervals, 0 - 10, 10 - 20, ... , 90 - 100 cm and then mixed to obtain 10 composite samples representing the soil profile for each site. The analyses on the profile samples were then used to estimate polynomial regressions on depth to obtain coefficients for orthogonal trends to 100 cm that might also be used as site variables as described above (9.7). Corresponding calculations were made with analyses to other depths to obtain alternative sets of profile trend coefficients and comparisons of these alternatives in regressions with the yield variables showed that trend coefficients to a depth of 50 cm gave the best relationships, presumably because plant growth was little affected by soil properties below 50 cm. Consequently these were used as site variables for the general model.
- Rainfall. Similar computations of trend coefficients were made with the data for rainfall received at each site during a constant set of 14 day periods, for the duration of the experiments and for various other time intervals. Comparisons of regressions with the yield variables showed that trend coefficients for the rainfall received in the 13 fortnightly periods from 3 June to 1 December gave the best relationships so that these were chosen as site variables for the general model.

- Sowing day. Site variable data were obtained by recording the sowing day for the establishment of each experiment, data values being the number of days to establishment from the beginning of the year.
- Dummies for soil and location. In addition dummy site variable values were obtained for each site for the variables *S1, S2, S3, S4, S5* for the location of the experiments on one of the five kinds of soil and for *D1, D2, D3, D4, D5* for location of the sites within one of the five agricultural districts of the region, each with values 1 or 0 as explained above for Table 9.4. For example experiments on soil S2 in district D1 had site variable values *S2* = 1, *S1* = *S3* = *S4* = *S5* = 0 and *D1* = 1, *D2* = *D3* = *D4* = *D5* = 0.

Stage 2: Regression studies

The very large amount of yield and site variable data produced by the 139 experiments can be used to develop many possible alternative regressions for the yield-site variable relationships based on various selections of the site variables, various mathematical transformations of the variables to provide for curved relationships and various products of the variables to provide for interaction effects. Comparisons of alternative regressions produced in such studies are made with tests of significance for the contributions by the variables in the presence of each other and with R_a^2 values (4.3) for the proportion of variance accounted for by the regressions. Such studies can easily become confusing because many of the variables are strongly correlated with each other so that alternative regressions can give similar tests of significance and similar R_a^2 values. Moreover the importance of some variables may only become apparent after allowance for the effects of others. The confusion can be largely avoided however by applying scientific knowledge about the nature of the variables and their relationships. Thus the primary basis for the choice of variables and form for a regression for any yield variable should be scientific knowledge about the nature of the relationships and not simply the statistical basis of that which gives the highest tests of significance and R_a^2 values, as facilitated for example by computer stepwise regression procedures. Accordingly the regression studies must be guided by a scientist, using his knowledge about the nature of the yield and site variables, their relationships to each other and their likely relative importance for the region represented by the data and not by a statistician, or worse by a computer, simply seeking statistically best relationships. Failure to recognize this need can produce selection bias in estimates obtained from the regressions, as described below in section 9.4, with estimates determined by particular configurations in the data values rather than by causal relationships.

A general strategy for the development of regressions for general models is:

1. to compute simple correlation coefficients between the yield variables and all of the site variables,
2. to estimate first regressions with the site variables that produce the most meaningful and significant correlations,
3. to develop forms for the regression relationships with data transformation and interaction terms (products between variables) based on knowledge about the nature of the relationship,
4. to add other site variables to the regressions that produce statistically significant improvements, and finally
5. to examine the regressions thus developed to determine whether the effects they represent are in accord with existing knowledge about the variables and do not simply represent a chance set of values for the data from which they have been estimated.

Nevertheless it should be recognized that even with the best of attempts to follow this strategy, the estimated regressions are still likely to be biased towards relationships that are peculiar to the data set from which they have been estimated, because of lack of sufficient scientific knowledge to define precisely the regressions that should be estimated from the data.

The development procedure and associated difficulties are illustrated with examples from the regression studies on the relationship between the $\overline{Y}$ variable and the site variables listed in Table 9.8. The much more extensive studies with the many site variables for this project together with other details of mainly local interest are described elsewhere (Colwell 1977, 1979; Colwell and Morton 1984).

$\overline{Y}$ regressions The yield variable $\overline{Y}$ gave many significant correlations with the site variables, the most significant being with a soil test for soil phosphorus in the surface soil (T_p), the mean of the water soluble nitrate nitrogen analyses in the soil profile (M_n), the reactive soil aluminium as measured by an ammonium oxalate extraction procedure (Al), the mean of the 13 fortnightly rainfall totals (M_{rain}) and some of the dummy variables for kind of soil and location within agricultural districts of the region represented by the experiments. First regression studies suggested by the correlations led to the regressions 1 and 2 in Table 9.9 in which the site variable for soil nitrogen, phosphorus, aluminium, rainfall, kind of soil and district all make significant ($p < 0.05$) contributions.

Table 9.8 Yield and site variables for the general model regressions in Table 9.9.

Variable	Description
	Yield variables
$\overline{Y}$	Mean yield
A_n	Response to N fertilizer
A_p	Response to P fertilizer
B_{np}	Linear x linear interaction
B_n	Curvature for response to N fertilizer
B_p	Curvature for response to P fertilizer
	Site variables, kind of soil
S1	Dummy (0, 1) variable for soil classification,
S2	" " " "
S3	" " " "
S4	" " " "
S5	" " " "
	Site variables, district
D1	Dummy (0, 1) variable for location in district
D2	" " "
D3	" " "
D4	" " "
D5	" " "
	Site variables, soil analysis 0-10 cm
T_p	Soil test for phosphorus
Al	Ammonium oxalate soluble aluminium
	Site variables, soil profile 0-50 cm
M_n	Mean for profile
L_n	Coefficient for linear trend with depth
	Other site variables
R	Total rain, 3 June to 1 December
W	Sowing day of year

The regression 1 in Table 9.9 provides an estimate of $\overline{Y}$ for level of yield such that $\overline{Y}$ increases with increase in the mean soil profile nitrogen analyses (M_n), the soil test analysis for phosphorus (T_p) on the surface soil and the mean fortnightly rainfall for the later part of

Table 9.9 Example regressions of $\bar{Y}$ on the site variables listed in Table 9.8.

No.	Regression	R_a^2 (%)
1	$\bar{Y} = 1063 - 784.6M_n^{-1} + 13.65T_p + 0.4528Al + 126.5M_{rain}$	35.2
2	$\bar{Y} = 1310 + 427.6S2 + 546.2S3 + 831.4S4 + 937.6S5$ $+ 156.1D2 + 224.1D3 + 399.9D4 + 457.6D5$	26.1
3	$\bar{Y} = 4593 - 476.1M_n^{-1} + 431.3\,L_n + 10.34R - 29.42W$ $- 2.665Al - 132.7S2 - 254.1S3 - 484.8S4$ $+ 25.99S5 + 1897D2 + 1967D3 + 446D4$ $+ 1714D5 + 0.0184WAl + (-343S2 - 585S3$ $- 1353S4 + 110S5)L_n - (7.73D2 + 7.92D3$ $+ 3.75D4 + 6.6D5)R$	58.8

the growing season (M_{rain}) as would be expected from knowledge about these variables. Moreover, a better relationship is obtained with the reciprocal transformation for the nitrogen variable (M_n^{-1}) in accord with the usual curved relationship expected between soil nutrient level and crop yield but not also with the soil phosphorus variable, possibly because the test values did not cover a sufficiently wide range. The relationship with soil aluminium is not so easily explained however because the positive coefficient value means that crop yield increases with increase in aluminium level rather than decreases as would be expected if there were some form of toxic effect on yield and in any case the aluminium analyses are all relatively low so that direct effects on crop production seem unlikely. Nevertheless since the relationship is highly significant ($p < 0.001$), the Al variable cannot simply be excluded from the regression on the basis that its effect is due to a chance aberration in the data set. Accordingly it is included but with the caution that the cause of the effect is not understood. The effect could be indirect for example, by effects of aluminium on some other important site variable(s) not included in the project.

The regression 2 was estimated because it provides estimates that are based entirely on the dummy variables for kind of soil and location. These variables can be easily identified without the need for costly laboratory analyses and consequently provide a very convenient though poorer alternative basis for estimation to the estimates possible with the regressions 1 and 3. Only four of the dummy variables are used for the five soils and four for the five districts because as explained above, estimates are obtained for the fifth variable (*S1* or *D1* with the present selection) by giving the other four variables the value of zero. As explained in section 9.2, the omitted soil or district serves as a reference and the coefficients estimate the difference in $\overline{Y}$ of the soils or districts relative to these references as illustrated by comparisons of the calculated values for $\overline{Y}$ in Table 9.10 with the values for *S1* and *D1*. The effects of soil and location estimated in this way represent the effects of factors that are associated with the classifications, such as soil physical properties associated with the kind of soil and the climate

Table 9.10 Estimates of yield level, $\overline{Y}$, for location on each of five kinds of soil in each of five districts, obtained from the regression 2 in Table 9.9. Estimates in brackets are obtained by extrapolation, there being no data for the combinations of soil and district.

District	Soil				
	S1	S2	S3	S4	S5
D1	(1310)	1738	1856	(2141)	2248
D2	1466	(1894)	2012	2298	(2404)
D3	(1534)	(1962)	2080	2366	2472
D4	(1710)	(2138)	2256	(2541)	(2648)
D5	(1768)	(2195)	2314	(2599)	2705

Table 9.11 Estimates of yield level, $\overline{Y}$, for location on each of five kinds of soil in each of five districts, obtained from the regression 3 in Table 9.9, with data mean values for the other site variables. Values are for comparison with those in Table 9.10. Estimates in brackets are by extrapolation .

District	Soil				
	S1	S2	S3	S4	S5
D1	(2149)	2124	2080	(2092)	2140
D2	2449	(2424)	2380	2392	(2440)
D3	(2477)	(2453)	2408	2420	2468
D4	(1821)	(1796)	1752	(1764)	(1812)
D5	(2498)	(2474)	2429	(2442)	2489

associated with the districts, rather than the effects of the variables themselves - for this reason they are termed *dummies*. For this same reason calculated values for soil and location effects on $\overline{Y}$ in Table 9.10 are very different from those given below in Table 9.11, where allowance is made for the effects of some of the site variables that are associated with the classifications represented by the dummies. Note incidentally that the regressions allow for some confounding of soils with districts, as for example the fact that *S1* only occurs in *D2* and *S2* only in *D1* (Table 9.7), and also for extrapolated estimations of $\overline{Y}$ values for soil and district combinations for which there are no data, as with the bracketed estimates in Tables 9.10 and 9.11.

The regression 3 in Table 9.9 was developed by further experimentation with additional variables and combinations of variables from regressions 1 and 2. Thus it was found that:

1. the addition of the variable L_n for linear trend in the nitrogen analyses down the soil profile gave a significant improvement, the positive coefficient value indicating that soil nitrogen effects on yield become relatively smaller with increase in depth,
2. the phosphorus soil test variable T_p does not make a significant contribution to the regression in the presence of the dummy variables so that it can be omitted,
3. the variable W for sowing day made a significant contribution to the regression such that $\overline{Y}$ decreases with delay in the time at which the experiments were established,
4. the effect of the linear distribution of nitrogen down the soil profile varied with kind of soil so that a significant improvement was obtained by the addition of *soil* × L_n interaction terms,
5. the effect of the rainfall variable varied amongst the districts so that a significant improvement was obtained from *district* × R interaction terms, and
6. a significant improvement was obtained by the addition of the product WAl for a sowing day × aluminium interaction effect, although the reason for this effect was not apparent.

The regression 3 was developed on this basis for use in a general model. Clearly however with such an extensive set of data, several other alternative regressions could be developed, each with a supporting rationale and statistical tests of significance. There is no simple procedure for defining the best regression under these circumstances and the final choice will depend very much on knowledge, experience and rather inevitably, on the personal opinions of the researcher.

The regression 3 can be used to calculate $\overline{Y}$ values for soils and districts corresponding to those described above in Table 9.10, but now

Table 9.12 General model with regressions for yield variables on the site variables of Table 9.8.

Regression
$\bar{Y} = 4593 - 476.1M_n^{-1} + 431.3\,L_n + 10.34R - 29.42W - 2.665AI$ $- 132.7S2 - 254.1S3 - 484.8S4 + 25.99S5 + 1897D2 + 1967D3$ $+ 446D4 + 1714D5 + 0.0184WAI + (-343S2 - 585S3 - 1353S4$ $+ 110S5)L_n + (-7.73D2 - 7.92D3 - 3.75D4 - 6.6D5)R$
$R_a^2 = 58.8\%$
$A_n = -5.609 + 19.66M_n^{-1} - 0.4332T_p + 0.07811R - 0.01290AI$ $- 2.545D2 - 11.51D3 - 23.05D4 + 36.89D5 + 0.1471M_n^{-1}AI$ $+ (-76.67D2 + 170.3D3 + 136.1D4 - 17.98D5)M_n^{-1}$ $+ (-0.04952D2 - 0.07975D3 + 0.07148D4 - 0.1749D5)R$
$R_a^2 = 58.7\%$
$A_p = 67.55 + 511.3T_p^{-1} - 2.89M_n - 36.38L_n + 0.2215R + 8.734D2$ $- 11.45D3 + 38.0D4 - 61.9D5 - 18.43S2 - 50.77S3 - 66.74S4$ $- 63.45S5 + (-420.6D2 + 618.4D3 - 961.5D4 + 849.4D5)T_p^{-1}$
$R_a^2 = 50.9\%$
$B_{np} = -0.135 - 1.299S2 - 0.509S3 - 6.759S4 - 1.863S5 + 3.53M_n^{-1}$ $+ (2.11S2 + 3.83S3 + 63.22S4 + 19.4S5)M_n^{-1}$
$R_a^2 = 16.4\%$
$B_n = -0.791 - 0.371D2 - 1.709D3 + 0.249D4 + 0.082D5 - 0.0931T$
$R_a^2 = 12.4\%$
$B_p = -13.07 + 12.52S2 + 8.481S3 + 1.038S4 + 8.861S5 - 77.64T^{-1}$
$R_a^2 = 23.9\%$

with adjustments for the effects of the other site variables in the regression. Thus the values in Table 9.11 are calculated from regression 3 using the mean values for all of the other site variables. When allowance for the effects of site variables are made in this way, the effects of kind of soil and district are very different and much smaller as can be seen by comparison of the tables.

L_n, L_p, etc. regressions

Corresponding regression studies were used to develop the regressions for the other yield variables, L_n, L_p, etc. using statistical tests of significance, local knowledge and experience for the choice of variables and form for the regressions. The set of regressions developed in this way and listed in Table 9.12 constitute a general soil fertility model for the region represented by the experiments.

Stage 3: Application

The model in Table 9.12 may be used to obtain estimates of yield and the effects of fertilizer applications with the same general procedure as that described in previous examples. Estimates of the yield variables are obtained from sets of values for the site variables and these are then used to calculate estimates of yield functions with the form

$$Y = b_0 + b_1 N^{\cdot 5} + b_2 P^{\cdot 5} + b_3 (NP)^{\cdot 5} + b_4 N + b_5 P$$

corresponding to that used to represent the results of the fertilizer experiments that were used to sample the soil fertility of the region. The functions thus estimated for sets of values for the site variable can then be used to calculate estimates of yields and optimal fertilizer rates. In order to avoid extrapolation errors with these estimates, values for the site variables should only be chosen within the range of values for the experimental data, both with respect to the range of values for each variable considered individually and less obviously, with respect to ranges of *combinations* of values for the different site variables in the experimental data. Because the present example is very general, covering the effects of many site variables, particular care is needed to avoid the latter type of extrapolation. A consequence is that the model cannot be used to prepare simple tables showing the effects of all combinations of the variable values on fertilizer requirements corresponding to the Table 9.6 of the previous example where there are only two site variables. The problems with extrapolated estimates are avoided if the experiments which provided the data for the model represent the range of soil fertility in the model region and if estimates are only made with site variable values for actual sites within that region. Nevertheless the possibility of sites occurring in the region with sets of site variable values outside the range of the data should be recognized. Example computations for the present model and of extrapolation effects are given in Appendix A.

9.4 Accuracy of estimates from general models

General models provide a means for estimating yield functions for sites in the model region from measured or expected values for the site variables and these estimated functions can then be used to calculate estimates of yields, profits and optimal fertilizer rates in the same way as if the functions had been obtained by actually carrying out experiments at those sites. Because the general model consists of regressions that have been estimated from the data provided by a series of experiments, such estimates should be, as already noted, about the best possible on the basis of the available data. It is difficult however to indicate precisely the accuracy of the estimates with standard errors or such because data values for the yield variables are correlated with each other and because allowance must be made for the effects of fertilizer rate. So although it is simple to apply standard procedures to calculate standard errors for regression estimates of the individual yield variables, it is difficult to calculate corresponding procedures for estimates of yield functions. Moreover since any such estimates of standard error would necessarily be based on estimates of error variance for the experimental conditions that provided the data for the estimation of the model and since these conditions may differ for other sites and in other years, this does not seem very important - in any case there are other, often much greater sources of inaccuracy that cannot be represented by standard errors, as now described. The important point is that although estimates may not prove to be very accurate for any particular site and crop, they will nevertheless be about the best possible on average for the range of conditions represented by the experiments, since the regressions on which they are based are statistical best estimates.

Extrapolation error

When a series of fertilizer experiments are carried out in a region they provide in effect a sampling of the soil fertility of the region and the adequacy of this sampling depends on the location and the growing conditions for the experiments. If the experiments are confined to certain kinds of soil and are carried out using particular kinds of agricultural practice, then they can only represent these soils and practices. Similarly if experiments are rejected because of damage by drought, floods, disease, weeds, etc. the experiments can only represent the rather idyllic conditions where crop production is not affected by such catastrophes. Thus the "region" actually represented by the experiments is better defined by the conditions and sites sampled by the experiments than by the geographic boundaries for the region in which they were located. Estimates based on extrapolations beyond these conditions may be inaccurate, the degree of inaccuracy increasing with the extent of the extrapolation. Clearly the greater the number of experiments, the more random their location, and the longer the period in which they were conducted, the more representative the model will be of a region and the less vulnerable the estimates will be to this type of error. Thus the example models might be expected to be reasonably representative because they are based on many experiments, 47 and 139 respectively, carried out in several successive years. Nevertheless the accuracy of any estimates must be judged on the basis of the range of the site variables, the range of values for these site variables and the range of combinations of these values that are represented by the models.

Omission bias

Many factors affect yield variables in a variable region and ideally the regression equations for general models would have site variables for all of these factors. This however will usually not be possible because of limited research resources and inadequacies in the procedures available for measuring the factors so that regression equations will have insufficient and inadequate regressor variables to account for the variations in the yield variables. Compromises are inevitable, using the available site variable data and the available knowledge to develop the regression equations. Regressions developed with such compromises are subject to *omission bias*[4] caused by the omission of regressor variables or the use of inadequate variables to represent the effects of all of the factors that affect the dependent variable in regressions.

Omission bias becomes very pronounced when regressor variables are highly correlated with the omitted variables but must also be expected, though to a lesser degree, in any regression where a variable that affects the dependent variable has been omitted. The bias is demonstrated with the artificial data in Table 9.13 which have been

---

[4] The mathematical nature of this bias is described by Draper and Smith (1981) and Miller (1990).

Table 9.13 Artificial data for crop yield *Y*, reactive aluminium.in the soil *Al* and a soil test for readily soluble phosphorus *T*,.chosen to demonstrate effects of omission bias.

Y	*Al*	*T*	*Y*	*Al*	*T*
3000	2	1.0	2000	10	6.5
3250	2	2.0	2100	10	7.0
3150	2	1.6	800	14	6.5
2000	6	1.8	900	14	7.0
2200	6	2.5	1400	14	9.0
2600	6	4.0	200	18	9.0
1500	10	4.5	300	18	9.5
1900	10	6.0	500	18	10.0

chosen to produce obvious biases in the simple regressions

$$Y = 3445.8 - 170.8Al, \quad r^2 = 94.4\% \tag{9.22}$$

and

$$Y = 3258.8 - 276.9T, \quad r^2 = 72.9\% \tag{9.23}$$

The regressions are illustrated in Figs 9.1 and 9.2 and represent the relationships estimated from the set of data supposedly obtained at 16 sites for crop yield *Y*, a soil test *T* for available phosphorus and a soil analysis *Al* for toxic levels of aluminium. From knowledge about the nature of the variables, the yield *Y* is expected to increase with increase in *T* (soil phosphorus) and to decrease with increase in *Al* (toxic aluminium). Since however the data values for *T* and *Al* are highly correlated ($r = 0.95$) there is a serious bias in each of the regression estimates of these relationships, even though both regressions are highly significant ($p<0.001$), because of omission in each of the other variable. The bias is seen by comparison of the regressions with the multiple regression

$$Y = 3363.4 - 301.0Al + 252.0T, \quad R^2 = 99.9\% \tag{9.24}$$

illustrated in Fig. 9.3, where provision is made for the simultaneous effects of *Al* and *T*. Thus the regression of *Y* on *Al* (9.22) shows the expected decrease in crop yield with increase in soil aluminium but the multiple regression (9.24) shows a much greater effect, as indicated by comparison of the respective regression coefficients for Al, –170.8 and –301.0. Even more obviously, the regression on *T* (9.23) shows an improbable but statistically highly significant decrease in yield with

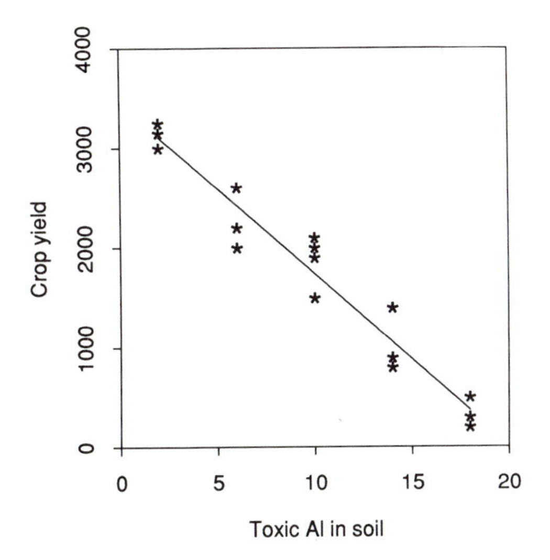

Fig. 9.1 Graph for the regression $Y = 3445.8 - 170.8Al$ estimated from the data of Table 9.13, showing a biased estimate of the relationship between crop yield and soil aluminium.

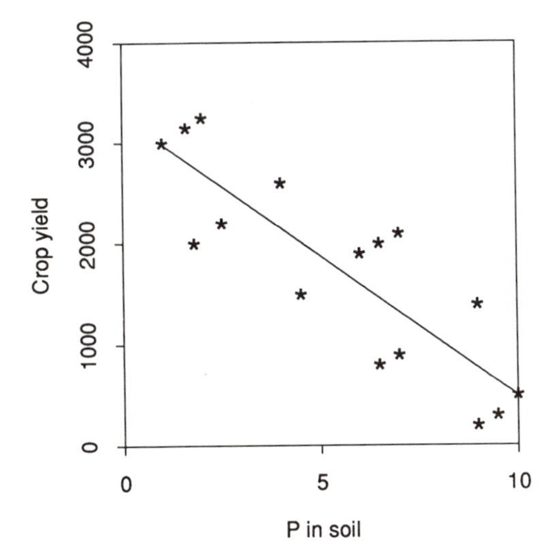

Fig. 9.2 Graph for the regression $Y = 3258.8 - 276.9T$ showing a biased estimate of the relationship between crop yield and soil phosphorus (T).

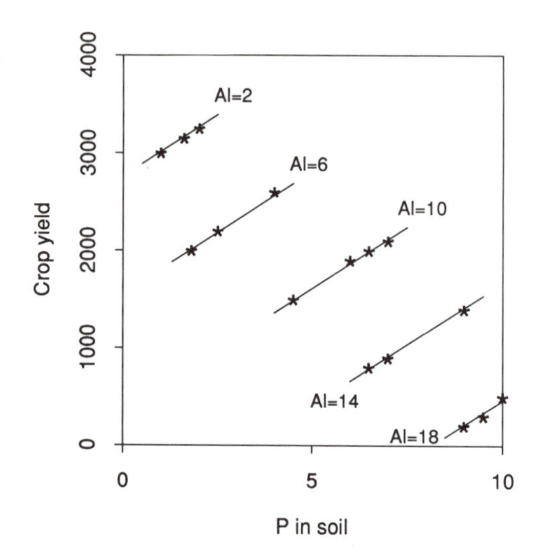

Fig. 9.3 Graph for the regression $Y = 3363.4 - 301.0Al + 252.0T$ showing an unbiased estimate of the effects of soil aluminium and phosphorus on crop yield.

increase in soil phosphorus with the term $-276.9T$ whereas the multiple regression shows the expected positive effect, with the term $+252.0T$.

Corresponding bias effects must be expected with the regressions for the example general models, particularly for that in Table 9.5 where there are only two site variables T and pH, given the many factors that must affect crop yields in the region represented by the experiments. With such limited data the researcher can only hope that the omitted variables vary more or less randomly and independently of the regressor variables. It is important therefore to develop regressions that include variables for as many factors as possible, and to examine estimated effects for conformity with scientific knowledge or previous experience. Indications of serious bias are large residual mean squares in regression analyses of variance and estimates that produce unlikely types of relationship.

Selection bias

A different and more subtle form of bias, called selection or competition bias (Miller 1984, 1990) is produced when the same data is used both to select regressor variables and to estimate the coefficients for these variables in regressions. Thus bias is likely to be produced

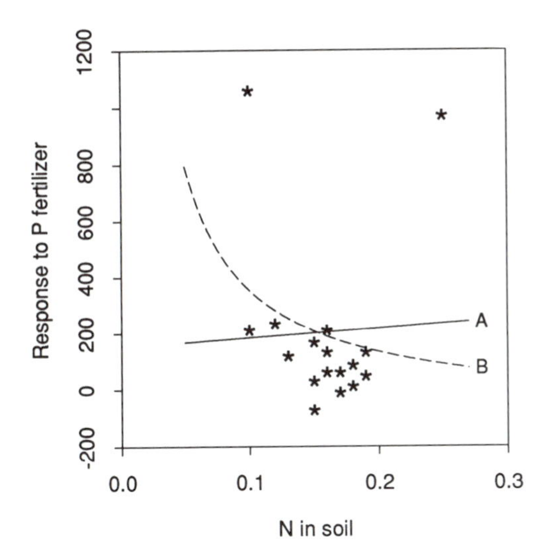

Fig. 9.4 Poor regressions between yield response to phosphorus fertilizer and soil nitrogen, curve A for (9.25) and B for (9.26).

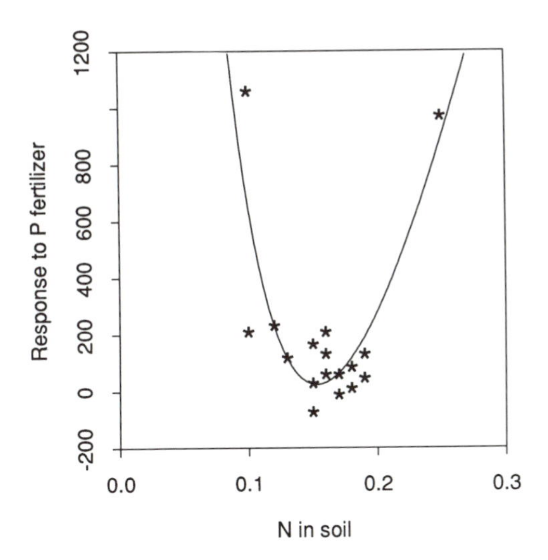

Fig. 9.5 Nonsense relationship due to selection bias between yield response to phosphorus fertilizer and soil analysis for nitrogen estimated by the regression (9.27).

when there is no theory or experience to define the model form for regression relationships so that models are chosen on the simple basis of obtaining good tests of significance. For example if a regression estimated with the model form

$$Y = b_0 + b_1 N + b_2 N^2$$

for the relationship between yield and fertilizer nutrient application rate gives a higher test of significance or a greater R_a^2 value (equation 8.12) than is obtained with a regression estimated from the same data with the alternative model form

$$Y = b_0 + b_1 N^{.5} + b_2 N$$

it may seem only sensible to use the former model. Since however experience with fertilizer experiments indicates that the second model gives a much more realistic representation of the relationship between Y and N, it should be chosen to estimate the true relationship, the higher significance and R_a^2 for the first model being attributed to selection bias associated with error aberrations amongst the data.

Selection bias can be produced very efficiently by the use of computing algorithms for the stepwise development of regressions that search for variables that produce "best" regressions as indicated by tests of significance or R_a^2 values. For example the following regressions were produced by a computer search for the most highly significant relationships between the results of a series of fertilizer experiments and soil analyses. Data for yield response of a particular crop to applications of nitrogen and phosphorus fertilizers had been obtained by experiments at 17 sites and first studies showed expected relationships between the yield variable A_p for yield response to P fertilizer and a soil test (analysis) T_p for readily soluble soil phosphorus. The studies also showed the expected absence of any apparent relationship of A_p to a soil test T_n for nitrogen as shown by negative R_a^2 values[5] for regressions on T_n and T_n^{-1}

$$A_p = 150.5 + 335.5 T_n \quad , \quad R_a^2 = -\ 6.5\% \tag{9.25}$$

$$A_p = -86.75 + 44.05 T_n^{-1} \quad , \quad R_a^2 = -1.4\% \tag{9.26}$$

and the graphs in Fig. 9.4. When however both of the regressor variables T_n and T_n^{-1} were included in the multiple regression

---

[5] R_a^2 is proportion of the variance accounted for by the regression (equation 4.3). A negative value occurs when the residual mean square in the analysis of variance is greater than the variance.

$$A_p = -6937.5 + 22791^{***} T_n + 531.8^{***} T_n^{-1} \ , \ R_a^2 = 65.7^{***}\% \qquad (9.27)$$

a highly significant relationship was indicated with the very high statistical probability rating *** (p<0.001) for a highly *improbable* relationship between response to phosphorus fertilizer and a soil test for nitrogen.

The subtlety of the selection bias hazard was shown by scientists whose first reaction on being shown the regression (9.27) was to attempt to produce an esoteric rationale to justify the acceptance of an obviously unlikely relationship, simply because it had a high level of statistical significance. The bias in this case was very obvious however, from the graph for the relationship in Fig. 9.5, illustrating that the computer had simply discovered a nonsense model form to represent a chance configuration in the data. As stressed already, the scientist rather than the statistician or the computer must ultimately be responsible for the choice of models to be estimated by the regression procedure.

Selection bias is likely to be produced when there is little or no knowledge concerning the true form of relationships between variables. In such cases the scientist has no alternative but to choose regression models that best represent relationships suggested by a particular set of data but should note nevertheless the possibility of selection bias and the need for further research to determine the true form for the relationship[6]. Thus the relationships between the yield variables B_n and B_p for curvature and the site variables in the example general model in Table 9.12 are likely to be affected by selection bias since there is no knowledge or experience to indicate the true nature of the relationships. In such cases, estimated regressions serve to represent relationships shown by a set of data rather than to estimate a known type of relationship.

Indirect relationships

A basic problem with the regression procedure is that it uses data to provide an estimate of a relationship between variables, in the form of a dependent variable expressed as a function of regressor variables, irrespective of whether the relationship is direct due to effects of the regressor variables on the dependent variable or indirect due to relationships of the dependent and regressor variables with some other variable or variables not included in the regression. When there are

---

[6] Bias may also be indicated by regression diagnostic procedures as described for example by Gunst and Mason(1980), Cook and Weisberg (1982) and Atkinson (1985).

indirect relationships, the regression may suggest a nonsense relationship, such as that illustrated above with Fig. 9.2, or give misleading impressions about its usefulness. For example, in some situations a highly significant regression relationship may be established between crop yield as a dependent variable and silicon uptake by crops as a regressor variable, suggesting that silicon increases crop yield and consequently that applications of silicon will increase yield whereas in fact the relationship is indirect representing the effect of water on crop growth, both yield and silicon uptake increasing with increase in water uptake by plants. Similarly there is sometimes confusion about the importance of soil classification because relationships between crop yield and classification units is due to indirect relationships with the soil factors that determine yield. For example if there is a relationship between the dummy variables *S1*, *S2* and *S3* for kind of soil and the variables *pH* and T_p for soil analyses for *pH* and phosphorus, the dummy variables and the analysis variables may substitute for each other in regressions to produce very different significance ratings for the contributions of these variables to regressions, as represented schematically with the following alternative regression models:

$$\overline{Y} = b_0 + b_1^{***}S2 + b_3^{***}S3 \tag{9.30}$$

$$\overline{Y} = b_0 + b_1^{***}pH + b_2^{***}T_p \tag{9.31}$$

$$\overline{Y} = b_0 + b_1S2 + b_2S3 + b_3pH + b_4^{*}T_p \tag{9.32}$$

For this example the regression (9.30) shows that kind of soil is closely related to $\overline{Y}$ for level of yield and similarly the regression (9.31) shows that $\overline{Y}$ is closely related to the soil analyses *pH* and T_p. These relationships are not apparent however in the regression (9.32) which shows no significant relationship with kind of soil or soil *pH* and only a weak relationship with the analysis T_p. Such confusion is resolved if an hierarchy of site variables can be established on the basis of scientific knowledge to indicate those that have the more direct causal effects on yield variables and hence those that should be selected first as site variables for regressions. For this example, scientific knowledge would suggest that yield is likely to be directly related to soil phosphorus as represented by the variable T_p , less directly with soil pH which often produces its effects by affecting the availability of soil nutrients or toxic effects of certain elements and less directly again with kind of soil where effects depend entirely on soil properties associated with the classification units. On this basis the following sequence of regressions would be estimated.

$$\overline{Y} = a_0 + a_1 T_p \quad (9.33)$$

$$\overline{Y} = b_0 + b_1 T_p + b_2 pH \quad (9.34)$$

$$\overline{Y} = c_0 + c_1 T_p + c_2 pH + c_3 S2 + c_4 S3 \quad (9.35)$$

Then regression (9.34) would be preferred only if the coefficient b_2 was significant showing that the contribution by pH was significant after allowing for the effect of T_p. Similarly the regression (9.35) would be preferred only if the variables S2 and S3 made a significant contribution to the regression after allowing for the effects of T_p and pH (in this case significant ratings for the coefficients are not appropriate because there are two variables involved). On this basis, the regression (9.31) on pH and T_p for the above example would be preferred even if the R_a^2 value for the regression (9.30) on soils was greater.

Weather uncertainties

The weather is obviously a major variable affecting the way crops grow and respond to fertilizers and also obviously, predictions of the weather are uncertain. Consequently, although site variables for the weather are important they serve to explain observed variations in the value of yield variables obtained in the past and can only be used to predict values for future crops on the basis of expected weather. Farming is carried out on this basis of course, the effects of bad years being compensated by those of good, so that model estimates for average weather conditions can be valuable as a long term guide for farming. Consequently, even if research resources do not allow the measurement of weather variables for their incorporation in general models, if the experiments from which the model has been derived cover a sufficiently representative range of weather conditions, the estimates will be for average conditions. With extensive sets of data more specific estimates may be possible for climatic districts within the region represented by the model, dummy variables being used to allow for differences between districts as in the example model in Table 9.12.

Estimates for different crop varieties or species

General models are developed from the results of a series of experiments, all of which have been conducted with the same crop species and variety, or cultivar. It seems inevitable however that the model will be used to obtain estimates for other varieties and species since it is not practicable to develop general models for every type of crop that is being grown or will be grown in a region. Although it may be reasonable to assume that model estimates, based on experiments with a particular crop variety, will also apply reasonably closely to similar varieties, the assumption will become less acceptable with time, given the constant stream of improved varieties being produced by plant breeders. In any case corresponding assumptions for different crop species are doubtful, as for example for the fertilizer requirements of barley from a general model for wheat. All such estimates are made in effect by extrapolation beyond the data from which the model was estimated and the greater the extrapolation the less reliable they must become. Relationships can be expected however between the way different crop species and varieties are affected by site variables or, more specifically, between the way yield variables representing the results of fertilizer experiments for different species and varieties relate to site variables. Bridging functions can thus be envisaged, relating yield variables for different kinds of crop in a region in such a way that yield models for a particular crop variety can be used to estimate corresponding models for other varieties and species. No studies seem to have been made however to investigate the feasibility of developing and using such functions. In the meantime estimates must be based on an assumed transferability of results and this introduces an additional potential source of error.

9.5 Alternatives to general models

The many sources of inaccuracy that reduce the accuracy of estimates from general models should not be used to denigrate the usefulness of the models. Consequently it is worth emphasizing that the same sources of error will affect corresponding estimates of crop yields and fertilizer effects obtained on any alternative basis. Also, in so far as estimates obtained from general models are based on regressions that have been estimated from a particular set of experimental data using existing scientific knowledge about the nature of the relationships, they must be better than estimates obtained from the same data by alternative procedures. If for example a model estimate proves to be less accurate than an estimate based on local farming experience, this must be due to variables covered by “experience” not being included in the general model and would indicate a need for further research.

Apart from such statistical considerations, the usefulness of general models may be judged by comparison with alternative procedures for obtaining corresponding estimates of crop yields and the effects of fertilizers for a region based on the results of a series of fertilizer experiments. Consider for example the following two alternative procedures that have been widely used in the past to apply the results of fertilizer experiments for the benefit of farmers in variable regions.

Alternative 1: Direct estimates by fertilizer experiments

Direct estimates of fertilizer-yield functions from the data of well conducted fertilizer experiments will be more accurate than estimates from general functions but *only* for the growing conditions that prevailed for the experiments. Thus the results obtained from a carefully conducted experiment on a research station, of the type favoured for publication, will be misleading if used to obtain estimates of crop yields and fertilizer requirements for other years and other sites because of the inevitable differences in growing conditions. General models are developed to provide a basis for allowing for the effects of site variables and accordingly it is not appropriate to compare the accuracy of estimates from individual experiments for very specific conditions with the estimates obtained from general models.

Alternative 2: Relative yield as an avoidance device

The obvious problem with estimating fertilizer-yield relationships for a region from the data of fertilizer experiments is that crop yields vary amongst different experiments due to variations in the many factors that affect plant growth and this of course is the reason for developing general models. An alternative procedure that has been very popular in the past is to express the data for each experiment on a relative yield or $\%Y$ basis, by the calculation

$$\%Y = \frac{Y}{A} \times 100 \tag{9.36}$$

where Y is yield data from an experiment and A is an estimate of the maximum attainable yield for that experiment, as indicated by the experiment data. The conversion of Y to $\%Y$ greatly reduces the variation amongst the data from different experiments and provides a simple procedure for estimating fertilizer-yield relationships on the percentage scale. There are however serious problems produced by the conversion:

1. $\%Y$ values do not provide a basis for estimating economic fertilizer rates. Farmers must justify the cost of fertilizer applications by the increases in actual yields and the economic returns they produce and not by their effects on relative yield, as stated very clearly a long time ago by Bondorff (1952). The conversion equation (9.36) can be used to calculate real yields for the calculation of economic rates given a value for A but of course values for A vary widely, this being the reason for the calculation of $\%Y$ values in the first place. Thus the original problem of variations in yield levels, represented

here by A, reappears as soon as any attempt is made to use $\%Y$ values to estimate actual yields or economic fertilizer rates.

2. The maximum attainable yield A required for the calculation of relative yields is often poorly defined by experimental data. In these cases A must be estimated by an extrapolation or a subjective procedure with ample opportunity for the introduction of personal bias towards a desired result.
3. The functional relationship between yield and fertilizer rate differs in basic form for different growing conditions and is not simply proportional to maximum yield. This can be seen by comparing fertilizer experiment data for the same soil and site in successive years with different weather conditions (Colwell 1985a, b; Colwell *et al.* 1988).
4. The calculation of $\%Y$ data can produce statistical bias. This statistical objection arises because dividing yields by A effectively weights error deviations by the inverse of the maximum yield.

The general model procedure avoids all of these problems by separating variations amongst the experimental data due to variations in levels of yield as represented by the yield variable $\overline{Y}$ from the variations due to variations in treatment effects as represented by the other yield variables. For most economic calculations relating to the use of fertilizers, level of yield is irrelevant. Thus the problem of variations in yield level amongst experiments which has led to the use of the relative yield device is avoided in the general model procedure, simply by the statistical separation of the variance component associated with this variable.

Appendix A

Example Estimations with a General Soil Fertility Model

A general soil fertility model is a statistical estimate of the relationships between the results of fertilizer experiments and site variables in a variable region and consequently it may be used to obtain statistical estimates of corresponding results of experiments from nominated values for the site variables. This appendix details the application of this procedure with the example model in Table 9.12 of chapter 9.

Derivation

The derivation of the example model involved the following steps:

1. Data obtained with a series of fertilizer experiments with wheat were used to estimate yield functions with the form

$$Y = b_0 + b_1 N^{\cdot 5} + b_2 P^{\cdot 5} + b_3 (NP)^{\cdot 5} + b_4 N + b_5 P$$

2. The yield functions were converted into the series of yield variables described in Table A.1 in order to obtain variables that correspond to orthogonal components of the functions, using the Table 8.7 formulas.
3. Data for the site variables in Table A.2 were obtained for each of the experiments.
4. The regressions in Table A.3 were developed for the relationships between these yield and site variables. These constitute the general model for the region represented by the experiments.

Application

The model regressions are used to estimate yield variables from values for site variables and hence to estimate yield functions which may then be used to calculate estimates of yields, optimal rates, etc.

Table A.1 Yield variables for the general model.

Variable	Description
$\overline{Y}$	Mean yield
A_n	Response to N fertilizer
A_p	Response to P fertilizer
B_{np}	Linear x linear interaction
B_n	Curvature for response to N fertilizer
B_p	Curvature for response to P fertilizer

Table A.2 Site variables for the general model.

	Site variables
	Kind of soil
S1	Dummy (0, 1) variable for soil classification,
S2	" " " "
S3	" " " "
S4	" " " "
S5	" " " "
	District
D1	Dummy (0, 1) variable for location in district
D2	" " "
D3	" " "
D4	" " "
D5	" " "
	Soil analysis 0-10 cm
T_p	Soil test for phosphorus
Al	Ammonium oxalate soluble aluminium
	Soil profile 0-50 cm
M_n	Mean water soluble N for profile
L_n	Coefficient for linear trend of N with depth
	Other site variables
R	Total rain, 3 June to 1 December
W	Sowing day of year

Table A.3 Regression equations, constituting the general model.

Regression	R_a^2 (%)
$\overline{Y} = 4593 - 476.1M_n^{-1} + 431.3\,L_n + 10.34R - 29.42W - 2.665AI - 132.7S2 - 254.1S3 - 484.8S4 + 25.99S5 + 1897D2 + 1967D3 + 446D4 + 1714D5 + 0.01836WAI + (-342.6S2 - 585.2S3 - 1353S4 + 110S5)L_n + (-7.725D2 - 7.924D3 - 3.745D4 - 6.6D5)R$	58.8
$A_n = -5.609 + 19.66M_n^{-1} - 0.4332T_p + 0.07811R - 0.01290AI - 2.545D2 - 11.51D3 - 23.05D4 + 36.89D5 + 0.1471M_n^{-1}AI + (-76.67D2 + 170.3D3 + 136.1D4 - 17.98D5)M_n^{-1} + (-0.04952D2 - 0.07975D3 + 0.07148D4 - 0.1749D5)R$	58.7
$A_p = 67.55 + 511.3T_p^{-1} - 2.89M_n - 36.38L_n + 0.2215R + 8.734D2 - 11.45D3 + 38.0D4 - 61.9D5 - 18.43S2 - 50.77S3 - 66.74S4 - 63.45S5 + (-420.6D2 + 618.4D3 - 961.5D4 + 849.4D5)T_p^{-1}$	50.9
$B_{np} = -0.135 - 1.299S2 - 0.509S3 - 6.759S4 - 1.863S5 + 3.53M_n^{-1} + (2.11S2 + 3.83S3 + 63.22S4 + 19.4S5)M_n^{-1}$	16.4
$B_n = -0.791 - 0.371D2 - 1.709D3 + 0.249D4 + 0.082D5 - 0.0931T$	12.4
$B_p = -13.07 + 12.52S2 + 8.481S3 + 1.038S4 + 8.861S5 - 77.64T^{-1}$	23.9

1. Values for each of the site variables are nominated, measured or estimated for sites in the region represented by the model.
2. Estimates of yield variable values are calculated for the sites by substituting the site variable values in the regressions. (Mean values for the yield variables are used when there are no significant regressions.)
3. The estimated yield variable values are used to calculate estimates of yield functions $Y = b_0 + b_1 N^{.5} + b_2 P^{.5} + b_3 (NP)^{.5} + b_4 N + b_5 P$, that is of functions with the same form as that used to represent the experimental results, using the conversion formulas of Table 8.12.
4. The yield functions are used to calculate estimates of yields, profits for various fertilizer application rates, optimal rates and so on for the sites, as described in chapters 2 and 3.

Example estimates

The example estimates of yield variable values in Table A.4 and the corresponding fertilizer-yield function coefficients in Table A.5 are for the site variable values in Table A.4. The examples have been chosen to illustrate both a wide range of contrasting estimates that might be obtained for sites in the region represented by the model and the meaningless type of estimate that can be produced by an extrapolation in the combination of a seemingly innocuous set of variable values.

Specifically, estimates $\hat{\bar{Y}}$, $\hat{A}_n$, ..., $\hat{B}_p$ of yield variables values are obtained by substituting the site variable values in the general model regressions. Estimates of the yield function

$$Y = b_0 + b_1 N^{.5} + b_2 P^{.5} + b_3 (NP)^{.5} + b_4 N + b_5 P$$

are then calculated by using the equations of Table 8.12 to calculate values for the coefficients, that is by:

$$b_5 = \hat{B}_p, \; b_4 = \hat{B}_n, \; b_3 = \hat{B}_{np}$$

$$b_2 = \hat{A}_p - b_3 \overline{N^{.5}} - b_5 h_p$$

$$b_1 = \hat{A}_n - b_3 \overline{P^{.5}} - b_4 h_n$$

$$b_0 = \hat{\bar{Y}} - b_1 \overline{N^{.5}} - b_2 \overline{P^{.5}} - b_3 \overline{NP^{.5}} - b_4 \overline{N} - b_5 \overline{P}$$

using the values $\overline{N^{.5}} = 5.704$, $\overline{N} = 47.675$, $\overline{P^{.5}} = 4.030$, $\overline{P} = 23.80$, $\overline{N^{.5}P^{.5}} = 22.99$, $h_n = 10.346$ and $h_p = 7.313$, derived from the experiment treatment rates 0, 22.4, 56.2, 112.1 kg N/ha and 0, 11.2, 28.0, 56.0 kg P/ha by the formulas of Tables 8.6 and 8.7. The estimated yield functions can then be used to calculate estimates of yield and profit for particular fertilizer applications, and of optimal fertilizer rates as described in chapters 2 and 3. This is illustrated graphically in Fig A.1 with estimates of yield that would be obtained with the N and P treatment rates of the original experiments and by the estimates of optimal fertilizer rates in Table A.7, using the example economic variable values V = \$0.08 / kg wheat, C_n = \$0.72 / kg N, C_p = \$1.92, E_n = 10.8 and E_p = 28.8.

The calculated estimates are reliable statistical estimates for sites A to D since the site variable values correspond to actual sets of data values from the data used to derive the model and consequently do not entail extrapolation. If site data are obtained from the region represented by the experiments corresponding estimates for other sites can similarly be expected to be reasonably reliable on the basis that they are not expected to involve estimation by extrapolation. Estimates with site variable values or combination of values outside the range of the experimental data will be less reliable however, possibly involving serious exprapolation errors. The need to consider *combinations* of site variable values as well as the magnitudes of the individual variables to avoid such extrapolation errors is illustrated by the example labelled "Extrapolation". The values for the variables T_p, Al, M_n, L_n for soil analyses, R for rainfall and W for time of sowing for this example are identical to those for Site D, but values for the dummy variable for kind of soil and district are different, producing an extrapolation, because the combination of values for T_p, Al, M_n, L_n, R and W do not occur for soil *S3* in district *D3*. The extrapolation produces an unlikely set of estimates as shown by the negative yields in Fig. A.1 and the very large optimal rates in Table A.7. Unfortunately such extrapolation effects are often not so obvious nor is extrapolation easily detected from a simple inspection of the range of values covered by the site variable data for the experiments from which the model is estimated. Serious extrapolation is unlikely, however, though still possible, if estimates are only made with site data values for sites in the region represented by the experiments, and if a sufficient number of experiments have been carried out to give an adequate sampling of the range of site variable values that occur in the region. In other words, general models can only be expected to provide reliable estimates if they have been derived from adequate sets of data and if estimates are confined to the region and conditions represented by the data. The problem of obtaining adequate data increases with the size of the region represented by a model and when many site variables affect the results of experiments so that in general, models are best developed for small regions.

Table A.4 Estimates of yield variable values from the regressions of Table A.3 with the site variable values of Table A.4.

Yield variable	Site A	Site B	Site C	Site D	Extrapolation
$\hat{Y}$	1661	3084	2487	1864	1603
$\hat{A}_n$	– 6.391	27.324	69.110	91.798	348.849
$\hat{A}_p$	162.173	286.983	131.349	189.107	197.466
$\hat{B}_{np}$	0.188	4.615	1.573	3.876	7.720
$\hat{B}_n$	– 2.158	– 3.031	– 3.878	– 2.382	– 3.720
$\hat{B}_p$	–20.326	– 25.653	– 9.835	– 18.997	– 10.516

Table A.5 Coefficients for $Y = b_0 + b_1N^{.5} + b_2P^{.5} + b_3(NP)^{.5} + b_4N + b_5P$ calculated from estimates of the yield variables in Table A.5.

Coef.	Site A	Site B	Site C	Site D	Extrapolation
b_0	909	1698	1500	532	– 1106
b_1	15.182	40.082	102.893	100.815	356.220
b_2	309.738	448.249	194.295	305.910	230.328
b_3	0.188	4.615	1.573	1.573	7.720
b_4	– 2.158	– 3.031	– 3.878	– 3.878	– 3.720
b_5	–20.326	–25.653	– 9.835	– 9.835	– 10.516

Table A.6 Example sets of values for the site variables of Table A.3. Values for Sites A to D correspond to data values at experiment sites. The values in the column "Extrapolation" represent a combination in the values for T_p , ... , W that do not occur amongst the data for soil *S3* in district *D3*.

Variable	Site A	Site B	Site C	Site D	Extrapolation
$S2$	0	0	0	0	0
$S3$	0	0	1	0	1
$S4$	0	1	0	0	0
$S5$	0	0	0	0	0
$D2$	1	0	0	1	0
$D3$	0	1	1	0	1
$D4$	0	0	0	0	0
$D5$	0	0	0	0	0
T_p	10.7	5.7	14.8	13.1	13.1
Al	580	1200	1140	1015	1015
M_n	10.94	5.8	3.32	0.88	0.88
L_n	− 0.592	− 0.247	− 0.217	− 0.034	− 0.034
R	395	484	232	484	484
W	214	150	145	244	244

Table A.7 Optimal nutrient application rates and wheat yields, yield response and profit increase for these applications, calculated from the estimated functions in Table A.5.

Site	Optimal rates		Yield	Yield response	Profit increase
	kg N/ha	kg P/ha	kg /ha	kg/ha	$/ha
A	0.4	9.9	1692	783	85.27
B	4.6	17.7	3244	1546	160.44
C	13.3	6.7	2276	775	65.99
D	18.7	11.4	1796	1264	111.22
Extrapolation	179.4	18.0	4225	5331	312.98

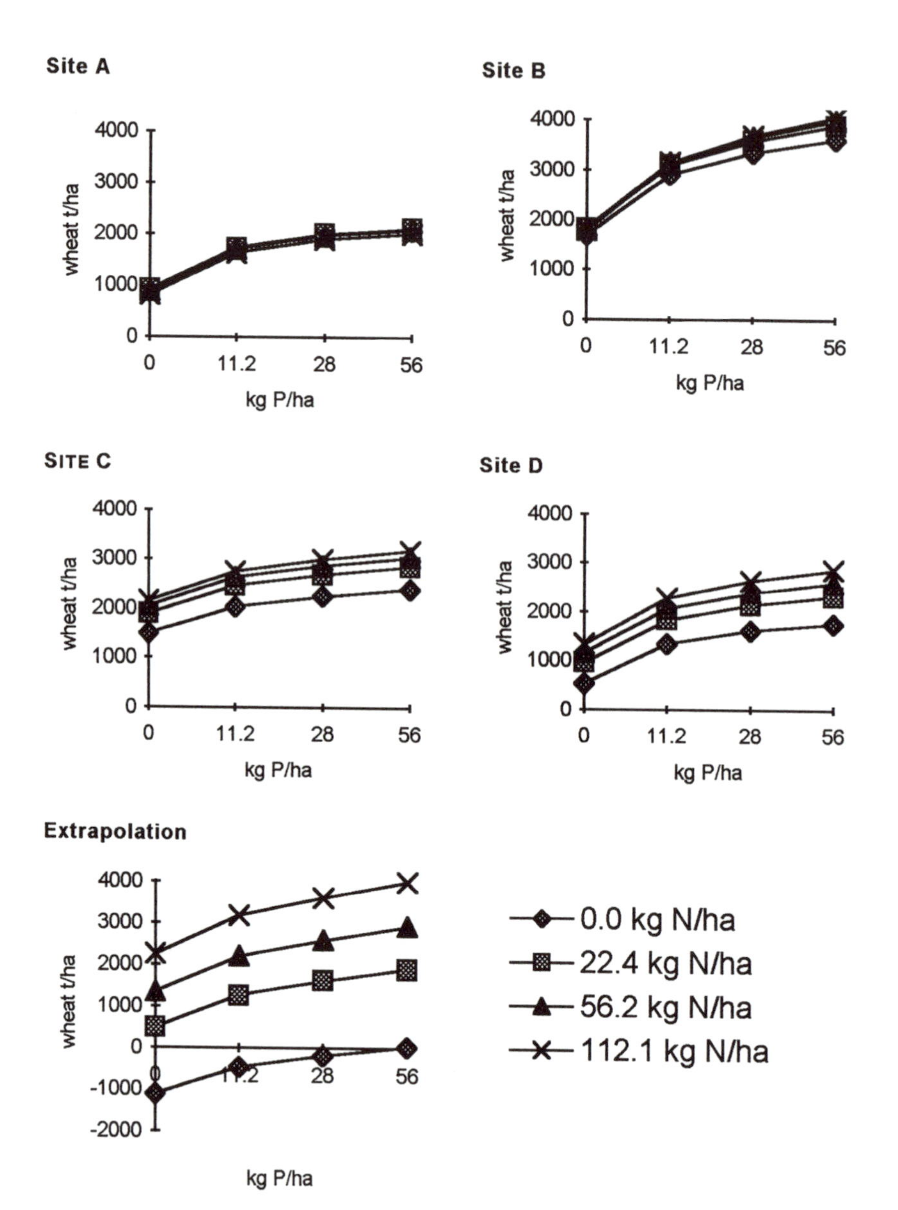

Fig. A.1 Graphical representation of yields estimated for the example sets of site variables with the example general model, for all combinations of the experimental treatment rates, 0, 22.4, 56.2, 112.1 kg N/ha and 0, 11.2, 28.0, 56.0 kg P/ha.

Appendix B

Identity of Regression and General Model Estimates

In chapter 8 general soil fertility models are defined as sets of regressions of yield variables on site variables that may be rearranged in the form of variable parameter equations corresponding in form to the fertilizer-yield functions that were used to develop the models. Because the variable parameter equations are derived by the combination of regression equations for yield variables that relate directly to the orthogonal polynomial components of fertilizer-yield functions, they are identical to direct regression estimates of the same variable parameter models, with the same experiment treatment rates and the same site variables as regressor variables. Corresponding derivations of general models with variables that do not meet this condition will not, in general, give equations that correspond with direct least squares regression estimates and the models will consequently be misleading.

The identity between estimates obtained with the set of regressions that constitute a general model and estimates obtained directly from experimental data with corresponding single regression is conveniently demonstrated by using the coefficients of the orthogonal polynomial regressions for fertilizer-yield functions rather than yield variables, to derive general relationships in the form of a variable parameter type of relationship. For example if a set of yield functions of the form $Y = b_0 + b_1N^{.5} + b_2N$ is converted to the orthogonal polynomial form $Y = p_0Z + p_1L + p_2Q$ as described in chapter 6, a generalization of the form $Y = p_0Z + p_1L + p_2Q$ can be obtained by establishing regression relationships of the coefficients on site variables

$$p_0 = f_0(r_1, r_2, \cdots, r_{k_a})$$
$$p_1 = f_1(s_1, s_2, \cdots, s_{k_b})$$
$$p_2 = f_2(t_1, t_2, \cdots, t_{k_c}) \quad \text{(B.1)}$$

with respectively k_a, k_b, and k_c regressor variables. Substituting the regressions for the parameters gives a generalization in the varying parameter form

$$Y = f_0(r_1, r_2, \cdots, r_{k_a})Z + f_1(s_1, s_2, \cdots, s_{k_b})L + f_2(t_1, t_2, \cdots, t_{k_c})Q \quad \text{(B.2)}$$

The equation obtained in this way by a combination of the regressions (B.1) is identical with that obtained by a direct regression estimate of (B.2) because the p_0, p_1, p_2 correspond to orthogonal components of analyses of variance of the regression $Y = p_0 Z + p_1 L + p_2 Q$ (or $Y = b_0 + b_1 N^{.5} + b_2 N$). Since the yield variables $\overline{Y}$, A_n and B_n relate directly to the trend coefficients with the constants g_{00}, g_{11} and g_{22} such that $\overline{Y} = g_{00} p_0$, $A_n = g_{11} p_1$ and $B_n = g_{22} p_2$ (equations 6.22, etc.), the same identity applies to generalizations with yield variables, such as the example equation (8.12) in chapter 8.

The identity is due essentially to the use of yield variables that relate directly to orthogonal components of regressions for the fertilizer-yield relationship, that is to the coefficients of polynomial regressions in an orthogonal form. Thus the yield variables relate to the coefficients p_i of orthogonal regressions for fertilizer-yield functions such as

$$Y = p_0 Z + p_1 L + p_2 Q, \; Y = p_0 Z + p_1 L_n + p_2 L_p + p_3 L_n L_p + p_4 Q_n + p_5 Q_p,$$

etc. which correspond to the non-orthogonal polynomials on the square root scale

$$Y = b_0 + b_1 N^{.5} + b_2 N, \quad Y = b_0 + b_1 N^{.5} + b_2 P^{.5} + b_3 (NP)^{.5} + b_4 N + b_5 P,$$

etc. as described in chapter 6. Regressions with orthogonal regressor variables have an important feature. Thus with the usual matrix notation, the regressions may be written in the form

$$\mathbf{y} = \mathbf{Xb} \quad \text{(B.3)}$$

where **y** is the data vector for the dependent variable, **X** is the matrix of regressor variables and **b** is the vector of regression coefficients, leading to the equation for the estimation of regression coefficients

$$\mathbf{b} = (\mathbf{X'X})^{-1}\mathbf{X'y} \quad \text{(B.4)}$$

In the special case when the matrix **X** is orthogonal, this estimation equation simplifies to

$$\mathbf{b} = \mathbf{X'y} \tag{B.5}$$

since then $(\mathbf{X'X}) = (\mathbf{X'X})^{-1} = \mathbf{I}$ where $\mathbf{I}$ is the identity matrix. Orthogonality between regressor variables has other important consequences for the generalization of regressions as now described for the orthogonal form $Y = p_0Z + p_1L + p_2Q$ of the simple yield function $Y = b_0 + b_1N^{\cdot 5} + b_2N$. Corresponding consequences can be demonstrated similarly with regressions for multiple nutrient functions $Y = p_0Z + p_1L_n + p_2L_p + p_3L_nL_p + p_4Q_n + p_5Q_p$, etc.

Given a series on n fertilizer experiments with the nutrient N, each with the same set of m treatments, coefficients for the orthogonal polynomial regression $Y_i = p_{0i}Z + p_{1i}L + p_{2i}Q$ are estimated by

$$\begin{bmatrix} p_0 \\ p_1 \\ p_2 \end{bmatrix}_i = \begin{bmatrix} z_1 & z_2 & \cdots & z_m \\ l_1 & l_2 & \cdots & l_m \\ q_1 & q_2 & \cdots & q_m \end{bmatrix} \begin{bmatrix} y_1 \\ y_2 \\ \vdots \\ y_m \end{bmatrix}_i \tag{B.6}$$

where the subscript i indicates the i th experiment, i = 1, 2. ... , n, $\begin{bmatrix} p_0 \\ p_1 \\ p_2 \end{bmatrix}_i$ is the vector of coefficients, $\begin{bmatrix} z_1 & z_2 & \cdots & z_m \\ l_1 & l_2 & \cdots & l_m \\ q_1 & q_2 & \cdots & q_m \end{bmatrix}$ is the matrix of orthogonal regressor variables for zero, linear and quadratic trends and $\begin{bmatrix} y_1 \\ y_2 \\ \vdots \\ y_m \end{bmatrix}_i$ is the vector of m yield data, corresponding respectively to $\mathbf{b}$, $\mathbf{X'}$ and $\mathbf{y}$ in the above equation $\mathbf{b} = \mathbf{X'y}$ for orthogonal regressions. If now the regression coefficients $\begin{bmatrix} p_0 \\ p_1 \\ p_2 \end{bmatrix}_i$ relate respectively to k_a, k_b, and k_c site variables, as represented by the regressions

$$\mathbf{p}_0 = \mathbf{Ra}, \ \mathbf{p}_1 = \mathbf{Sb}, \ \mathbf{p}_2 = \mathbf{Tc} \tag{B.7}$$

where the data matrices **R, S, T** have dimensions $(n \times k_a)$, $(n \times k_b)$ and $(n \times k_c)$, then the regression coefficients are estimated by

$$\mathbf{a} = (\mathbf{R}'\mathbf{R})^{-1}\mathbf{R}\mathbf{p}_0, \ \mathbf{b} = (\mathbf{S}'\mathbf{S})^{-1}\mathbf{S}\mathbf{p}_1, \ \mathbf{c} = (\mathbf{T}'\mathbf{T})^{-1}\mathbf{T}\mathbf{p}_2 \tag{B.8}$$

The yield function $Y = p_0Z + p_1L + p_2Q$ can then be generalized in a variable parameter form by substituting the regressions **Ra**, **Sb** and **Tc** for the parameters p_0, p_1 and p_2, that is by

$$\mathbf{y} = \mathbf{RaZ} + \mathbf{SbL} + \mathbf{TcQ} \tag{B.9}$$

The generalization obtained by substituting regressions for the coefficients in this way is identical to that obtained directly by fitting a large single regression to the combined data from all experiments with an appropriate set of regressor variables, because the coefficients p_0, p_1 and p_2 are for orthogonal variables. This can be shown by first arranging the fertilizer experiment data in a form suitable for the estimation of trend coefficients p_0, p_1 and p_2 for the n experiments in a single large regression and then defining a regression of these coefficients on the site variables. Thus if y_{ij} is the yield for the i th treatment in the j th experiment, $i = 1, 2, \ldots, m$ and $j = 1, 2, \ldots, n$ and p_{0j}, p_{1j}, p_{2j} are coefficient values for the j th experiment, then the regression is

$$\begin{bmatrix} y_{11} \\ y_{12} \\ \vdots \\ y_{1n} \\ y_{21} \\ y_{22} \\ \vdots \\ y_{2n} \\ \vdots \\ \vdots \\ y_{m1} \\ y_{m2} \\ \vdots \\ y_{mn} \end{bmatrix} = \begin{bmatrix} z_1 & 0 & \cdots & 0 & l_1 & 0 & \cdots & 0 & \cdots & \cdots & q_1 & 0 & \cdots & 0 \\ 0 & z_1 & \cdots & 0 & 0 & l_1 & \cdots & 0 & \cdots & \cdots & 0 & q_1 & \cdots & 0 \\ & & \ddots & & & & & & & & & & \ddots & \\ 0 & 0 & \cdots & z_1 & 0 & 0 & \cdots & l_1 & \cdots & \cdots & 0 & 0 & \cdots & q_1 \\ z_2 & 0 & \cdots & 0 & l_2 & 0 & \cdots & 0 & \cdots & \cdots & q_2 & 0 & \cdots & 0 \\ 0 & z_2 & \cdots & 0 & 0 & l_2 & \cdots & 0 & \cdots & \cdots & 0 & q_2 & \cdots & 0 \\ & & \ddots & & & & \ddots & & & & & & \ddots & \\ 0 & 0 & \cdots & z_2 & 0 & 0 & \cdots & l_2 & \cdots & \cdots & 0 & 0 & \cdots & q_2 \\ \vdots & & & & & & & & & & & & & \\ z_m & 0 & \cdots & 0 & l_m & 0 & \cdots & 0 & \cdots & \cdots & q_m & 0 & \cdots & 0 \\ 0 & z_m & \cdots & 0 & 0 & l_m & \cdots & 0 & \cdots & \cdots & 0 & q_m & \cdots & 0 \\ \vdots & & \ddots & & & & \ddots & & & & & & \ddots & \\ 0 & 0 & \cdots & z_m & 0 & 0 & \cdots & l_m & \cdots & \cdots & 0 & 0 & \cdots & q_m \end{bmatrix} \begin{bmatrix} p_{01} \\ p_{02} \\ \vdots \\ p_{0n} \\ p_{11} \\ p_{12} \\ \vdots \\ p_{1n} \\ p_{21} \\ p_{22} \\ \vdots \\ p_{2n} \end{bmatrix} \quad \text{(B.10)}$$

$$(mn \times 1) \qquad (mn \times 3n) \qquad (3n \times 1)$$

For convenience the large vectors and matrix of this equation may be written

$$\mathbf{y} = \mathbf{E} \begin{bmatrix} \mathbf{p}_0 \\ \mathbf{p}_1 \\ \mathbf{p}_2 \end{bmatrix} \quad \text{(B.11)}$$

where $\mathbf{y}$ is the $(nm \times 1)$ vector of the yield data, $\mathbf{E}$ is a $(nm \times 3n)$ matrix composed of orthogonal polynomial trends corresponding to treatment effects and $\mathbf{p}_0$, $\mathbf{p}_1$ and $\mathbf{p}_2$ are $(n \times 1)$ vectors of the p_0 , p_1 and p_2 trend coefficients for the n experiments. Since $\mathbf{E}$ is orthogonal, the vectors of trend coefficients for the n experiments are estimated simply by

$$\begin{bmatrix} \mathbf{p}_0 \\ \mathbf{p}_1 \\ \mathbf{p}_2 \end{bmatrix} = \mathbf{E}'\mathbf{y} \quad \text{(B.12)}$$

These estimates, obtained from the combination of all the experimental data in a large regression, are identical to those obtained by the n individual regressions (B.6), with the data from each of the n experiments. The important distinction between the alternative procedures, (B.6) and (B.12), is that standard analyses of variance and tests of significance can be carried out with the estimates for individual experiments but not with the estimates from the combined data, because error variances are different for each experiment.

Relationships corresponding to (B.7), between the n values for the p_0, p_1 and p_2 coefficients and the k_a, k_b, and k_c site variables can be estimated similarly with a single large regression combination of data from the n experiments, by constructing an appropriate $(3n \times [k_a + k_b + k_c])$ data matrix

$$\begin{bmatrix} p_{01} \\ p_{02} \\ \vdots \\ p_{0n} \\ p_{11} \\ p_{12} \\ \vdots \\ p_{1n} \\ p_{21} \\ p_{22} \\ \vdots \\ p_{2n} \end{bmatrix} = \begin{bmatrix} r_{11} & r_{12} & \cdots & r_{1k_a} & 0 & 0 & \cdots & 0 & 0 & 0 & \cdots & 0 \\ r_{21} & r_{22} & \cdots & r_{2k_a} & 0 & 0 & \cdots & 0 & 0 & 0 & \cdots & 0 \\ & & \ddots & & & & & & & & & \\ r_{n1} & r_{n2} & \cdots & r_{nk_a} & 0 & 0 & \cdots & 0 & 0 & 0 & \cdots & 0 \\ 0 & 0 & \cdots & 0 & s_{11} & s_{12} & \cdots & s_{1k_b} & 0 & 0 & \cdots & 0 \\ 0 & 0 & \cdots & 0 & s_{21} & s_{22} & \cdots & s_{2k_b} & 0 & 0 & \cdots & 0 \\ & & & & & & \ddots & & & & & \\ 0 & 0 & \cdots & 0 & s_{n1} & s_{n2} & \cdots & s_{nk_b} & 0 & 0 & \cdots & 0 \\ 0 & 0 & \cdots & 0 & 0 & 0 & \cdots & 0 & t_{11} & t_{12} & \cdots & t_{1k_c} \\ 0 & 0 & \cdots & 0 & 0 & 0 & \cdots & 0 & t_{21} & t_{22} & \cdots & t_{2k_c} \\ & & & & & & & & & & \ddots & \\ 0 & 0 & \cdots & 0 & 0 & 0 & \cdots & 0 & t_{n1} & t_{n2} & \cdots & t_{nk_c} \end{bmatrix} \begin{bmatrix} a_1 \\ a_2 \\ \vdots \\ a_{k_a} \\ b_1 \\ b_2 \\ \vdots \\ b_{k_b} \\ c_1 \\ c_2 \\ \vdots \\ c_{k_c} \end{bmatrix} \qquad \text{(B.13)}$$

$(3n \times 1) \qquad (3n \times [k_a + k_b + k_c]) \qquad ([k_a + k_b + k_c] \times 1)$

For convenience this regression can be written

$$\begin{bmatrix} \mathbf{p_0} \\ \mathbf{p_1} \\ \mathbf{p_2} \end{bmatrix} = \begin{bmatrix} \mathbf{R} & \mathbf{0} & \mathbf{0} \\ \mathbf{0} & \mathbf{S} & \mathbf{0} \\ \mathbf{0} & \mathbf{0} & \mathbf{T} \end{bmatrix} \begin{bmatrix} \mathbf{a} \\ \mathbf{b} \\ \mathbf{c} \end{bmatrix} \qquad \text{(B.14)}$$

and the coefficients estimated by

$$\begin{bmatrix} \mathbf{a} \\ \mathbf{b} \\ \mathbf{c} \end{bmatrix} = \begin{bmatrix} (\mathbf{R}'\mathbf{R})^{-1}\mathbf{R}'\mathbf{p}_0 \\ (\mathbf{S}'\mathbf{S})^{-1}\mathbf{S}'\mathbf{p}_1 \\ (\mathbf{T}'\mathbf{T})^{-1}\mathbf{T}'\mathbf{p}_2 \end{bmatrix} \tag{B.15}$$

Alternatively, by substituting $\mathbf{E}'\mathbf{y}$ for $\begin{bmatrix} \mathbf{y}_0 \\ \mathbf{y}_1 \\ \mathbf{y}_2 \end{bmatrix}$ from (B.12), equation (B.14) may be written as a regression of the yield data

$$\mathbf{y} = \mathbf{E}\begin{bmatrix} \mathbf{R} & \mathbf{0} & \mathbf{0} \\ \mathbf{0} & \mathbf{S} & \mathbf{0} \\ \mathbf{0} & \mathbf{0} & \mathbf{T} \end{bmatrix}\begin{bmatrix} \mathbf{a} \\ \mathbf{b} \\ \mathbf{c} \end{bmatrix} \tag{B.16}$$

giving the estimation equation now with the yield vector $\mathbf{y}$

$$\begin{bmatrix} \mathbf{a} \\ \mathbf{b} \\ \mathbf{c} \end{bmatrix} = \begin{bmatrix} \mathbf{R}^{-1}\mathbf{R} \\ \mathbf{S}^{-1}\mathbf{S} \\ \mathbf{T}^{-1}\mathbf{T} \end{bmatrix}\begin{bmatrix} \mathbf{R} & \mathbf{0} & \mathbf{0} \\ \mathbf{0} & \mathbf{S} & \mathbf{0} \\ \mathbf{0} & \mathbf{0} & \mathbf{T} \end{bmatrix}\mathbf{E}'\mathbf{y} \tag{B.17}$$

The estimate of **a**, **b** and **c** from the combined data, either by (B.15) or (B.17), is identical with the estimate from the individual regressions for p_0, p_1 and p_2 in (B.8). Thus given corresponding regressor variables as defined above by $\mathbf{E}\begin{bmatrix} \mathbf{R} & \mathbf{0} & \mathbf{0} \\ \mathbf{0} & \mathbf{S} & \mathbf{0} \\ \mathbf{0} & \mathbf{0} & \mathbf{T} \end{bmatrix}$, identical estimates of general models are obtained with regressions for yield variables, since these relate directly to trend coefficients, or by a direct estimate with a large multiple regression as demonstrated with an example in chapter 8. It is however much simpler to develop models with separate regressions for the yield variables as explained in chapter 8 because then the regressions may be based on scientific knowledge about relationships with site variables and because then standard tests of significance can be used.

References

Atkinson, A.C. 1985. *Plots, Transformations and Regression: An introduction to graphical methods of diagnostic regression analysis.* Clarendon Press, Oxford.

Balmukand, B.H. 1928. Studies on crop variation. V. The relationship between yield and soil nutrients. *J. Agric. Sci.* **18**, 602-627.

Bartlett, M.S. 1938. The approximate recovery of information from replicated field experiments with large blocks. *J. Agric. Sci.* **28**, 418-427.

Bondorff, K.A. 1952. The evaluation of soil analysis. *Trans. Int. Soc. Soil Sci. Dublin*, 290-295.

Box, G.E.P. and Wilson, K.P. 1951. On the experimental attainment of optimum conditions. *J. Roy. Statist. Soc., B,* **13**, 1-45.

Box, G.E.P. and Draper, N.R. 1959. A basis for the selection of a response surface design. *J. Amer. Statist. Assoc.*, **54**, 622-654.

Cochran, W.D. and Cox, G.M. 1957. *Experimental Designs.* John Wiley, New York.

Colwell, J.D. 1963. The estimation of the phosphorus fertilizer requirements of wheat in southern New South Wales by soil analysis. *Aust. J. Exper. Agric. Anim. Husb.*, **3**, 190-197.

Colwell, J.D. 1967. Calibration and assessment of soil tests for estimating fertilizer requirements. I. Statistical models and tests of significance. *Aust. J. Soil Res.,* **5,** 275-293.

Colwell, J.D. 1968. Calibration and assessment of soil tests for estimating fertilizer requirements. II. Fertilizer requirements and an evaluation of soil testing. *Aust. J. Soil Res.,* **6,** 93-103.

Colwell, J.D. 1970. A statistical-chemical characterization of four great soil groups in southern New South Wales based on orthogonal polynomials, *Aust. J. Soil Res.*, **8**, 221-238.

Colwell, J.D. 1977. *National Soil Fertility Project. Volume 1, Objectives and Procedures.* Commonwealth Scientific and Industrial Research Organization, Division of Soils, Adelaide, Australia.

Colwell, J.D. 1978. *Computations for Studies of Soil Fertility and Fertilizer Requirements.* Commonwealth Agricultural Bureaux, Farnham Royal, Slough, UK and Commonwealth Scientific and Industrial Research Organization, Melbourne, Australia, 297pp.

Colwell, J.D. 1979. *National Soil Fertility Project. Volume 2, Soil Fertility Relationships.* Commonwealth Scientific and Industrial Research Organization, Division of Soils, Adelaide, Australia.

Colwell, J.D. 1983. *Fertilizer Requirements.* In *Soils: An Australian Viewpoint.* CSIRO, Melbourne, Australia, and Academic Press, London, UK: 795-815.

Colwell, J.D. 1985a. Fertilizing Programs. 1. Variability in responses of successive crops to fresh and previous applications of phosphorus fertilizers, in Australia and Brazil. *Fertilizer Research,* **8**, 21-38.

Colwell, J.D. 1985b. Fertilizing Programs. 2. Optimal programs of fertilizer application. *Fertilizer Research,* **8**, 39-47.

Colwell, J.D. and Donnelly, J.D. 1971. Effects of soil composition on the relationship between soil test values for phosphorus fertilizer requirements. *Aust. J. Soil Res.*, **9**, 43-54.

Colwell, J.D. and Esdaile, R.J. 1968. The calibration, interpretation and evaluation of tests for the phosphorus fertilizer requirements of wheat in northern New South Wales. *Aust. J. Soil Res.*, **6**, 105-120.

Colwell, J.D. and Goedert, W.J. 1988. Substitution rates as measures of the relative effectiveness of alternative phosphorus fertilizers. *Fertilizer Research,* **15**, 163-172.

Colwell, J.D. and Morton, R. 1984. General or transfer models of relationships between wheat yields and fertilizer rates in southern Australia and statistical bias. *Aust. J. Soil Res.,* **22**, 191-205.

Colwell, J.D. and Stackhouse, K. 1970. Some problems in the estimation of simultaneous fertilizer requirements of crops from response surfaces. *Aust. J. Exper. Agric. Anim. Husb.*, **10**, 183-195.

Colwell, J.D., Suhet, A.R. and van Raij, B. 1988. *Statistical Procedures for Developing General Soil Fertility Models for Variable Regions.* Commonwealth Scientific and Industrial Research Organization, Division of Soils Report 93, Adelaide, Australia, 68pp.

Cook, R. and Weisberg, S. 1982. *Residuals and Influence in Regression.* Chapman and Hall, New York.

Draper, N.R. and Smith, H. 1981. *Applied Regression Analysis.* 2nd Edition. John Wiley, New York.

Fisher, R.A. 1924. III. The influence of rainfall on the yield of wheat at Rothamsted. *Philos. Trans.* B231, 89.

Fisher, R. A. 1937. *Statistical Methods for Research Workers.* Oliver and Boyd, Edinburgh.

Fisher, R.A. and Yates, F. 1963. *Statistical Tables for Biological Agricultural and Medical Research.* Oliver and Boyd, Edinburgh.

France, J. and Thornley, J.H.M. 1984. *Mathematical Models in Agriculture.* Butterworths, London.

Gunst, R.F. and Mason, R.L. 1980. *Regression Analysis and its Application.* Marcel Dekker, New York.

Hallsworth, E.G. 1969. The measurement of soil fertility: the national soil fertility project. *J. Aust. Inst. Agric. Sci.*, **35**, 78-89.

Hill, W.J. and Hunter, W.G. 1966. A review of response surface methodology: A literature survey. *Technometrics*, **8**, 571-590.

John, J.A. and Quenouille, M.H. 1977. *Experiments: Design and Analysis.* Charles Griffin, London.

Kempton, R.A. and Howes, C.W. 1981. The use of neighbouring plot values in the analysis of variety trials. *Appl. Statist.*, **30**, 59-70.

Khuri, A.I. and Cornell, J.A. 1987. *Response Surfaces: Designs and Analyses.* Marcel Dekker, New York.

Maddala, G.S. 1977. *Econometrics.* McGraw-Hill, New York.

Mead, R. 1988. *The Design of Experiments: Statistical Principles for Practical Applications.* Cambridge University Press, Cambridge.

Mead, R. and Pike, D.J. 1975. A review of response surface methodology from a biometric point of view. *Biometrics*, **31**, 803-851.

Miller, A.J. 1984. Selection of subsets of regression variables. *J. Royal Statistical Soc.*, **147**, 389-425.

Miller, A.J. 1990. *Subset Selection in Regression.* Chapman and Hall, London.

Mitscherlich, E.O. 1909. Das Gesetz des Minimums und das Gestez des abnehmenden Bodenertrages. *Landw. Jahrb.*, **38**, 537-552.

Nelder, J.A. 1966. Inverse polynomials, a useful group of multi-factor response functions. *Biometrics*, **22**, 128-141.

Papadakis, J.S. 1937. Méthode statistique pour les expériences cour champ. *Bull. Inst. Amél. Plants à Salonique*, No. 23.

Searle, S.R. 1966. *Matrix Algebra for the Biological Sciences.* John Wiley, New York.

Snedecor, G.W.and Cochran,W.G. 1967. *Statistical Methods.* Iowa University Press, Ames.

Sparrow, P.E. 1979. Nitrogen response curves of spring barley. *J. Agric. Sci. (Camb.)*, **92**, 307-317.

Sylvester-Bradley, R., Dampney, P.M.R. and Murray, A.W.A. 1984. The response of winter wheat to nitrogen. In *The Nitrogen Requirements of Cereals* (Eds P. Needham, J.R. Archer, R. Sylvester-Bradley and G. Goodlass), pp. 151-176, HMSO, London.

Tintner, G. and Millham, C.B. 1969. *Mathematics and Statistics for Economists.* Holt, Rinehart & Winston, New York.

Weisberg, S. 1980. *Applied Linear Regression*. John Wiley, NewYork.

Williams, E.R. 1986. A neighbour model for field experiments. *Biometrika,* **73**, 279-287.

Index